THE EIGEN THEORY OF THE PHYSICAL WORLD

Third Edition

Deepal S. Benaragama Ph. D.

Grosvenor House
Publishing Limited

All rights reserved
Copyright © Deepal Benaragama, 2022
Third Edition

The right of Deepal Benaragama to be identified as the author of this work has been asserted in accordance with Section 78 of the Copyright, Designs and Patents Act 1988

The book cover is copyright to Deepal Benaragama

This book is published by
Grosvenor House Publishing Ltd
Link House
140 The Broadway, Tolworth, Surrey, KT6 7HT.
www.grosvenorhousepublishing.co.uk

This book is sold subject to the conditions that it shall not, by way of trade or otherwise, be lent, resold, hired out or otherwise circulated without the author's or publisher's prior consent in any form of binding or cover other than that in which it is published and
without a similar condition including this condition being imposed on the subsequent purchaser.

A CIP record for this book
is available from the British Library

First Edition published by Grosvenor House Publishing
Paperback ISBN 978-1-83975-133-2
eBook ISBN 978-1-83975-451-7

Second Edition published by BP International
Paperback ISBN 978-9-35547-393-6
eBook ISBN 978-9-35547-394-3

Third Edition published by Grosvenor House Publishing
Paperback ISBN 978-1-80381-081-2
eBook ISBN 978-1-80381-121-5

To My Wife Lalitha

CONTENTS

PREFACE ix
ACKNOWLEDGEMENTS xi

CHAPTER 1 1
Introduction 1
1.1 The state of balance between point and line 2
1.2 The fundamental structure of the physical world
     as the state of balance between particle and field 5
1.3 A simple view of the field structure of a particle
     of matter 8

CHAPTER 2 11
Balance 11
2.1 Symmetry misinterpreted as balance 11
2.2 Balance as the highest principle of nature 12

CHAPTER 3 19
A brief History of Physics 19
3.1 The early stages of physics 19
3.2 Newtonian mechanics 28
3.3 Relativistic physics 33
3.4 Quantum mechanics 37

CHAPTER 4 39
The Eigen Theory of the Physical World 39
4.1 Introduction 39
4.2 A comprehensive introduction to the three stages
 of the Eigen Theory 40
  4.2.1 Wave aspect of the wave-particle duality
   of matter 40
  4.2.2 Particle aspect of the wave-particle duality
   of matter 45
  4.2.3 The basis of the physical world as a pair of
   balanced particles in the base manifold 49
  4.2.4 ET and the theories of Newton and Einstein 50
4.3 Stage 1 of the Eigen Theory 51
  4.3.1 Wave aspect of the wave-particle duality
   of matter 51
  4.3.2 The absolute conservation principle 53
  4.3.3 The transport equations of (v, u) and the field
   equation satisfied by (g, h) 55
  4.3.4 T-R unity of the wave aspect of matter and T-R
   duality of the corresponding field 57
4.4 Stage 2 of the Eigen Theory 58
  4.4.1 Particle aspect of the wave-particle duality
   of matter 58
4.5 Stage 3 of the Eigen Theory 60
  4.5.1 The base manifold 60
  4.5.2 The complete structure of the two particles 62
  4.5.3 The spherically symmetric cosmic central body
   system 63

CHAPTER 5	65
Simple applications of ET	65
5.1 The principle of uncertainty	65
5.2 Matter-antimatter	65
5.3 The double-slit experiment	66
5.3.1 Particles of matter	66
5.3.2 Photons	66
5.4 ET and Newtonian Mechanics	67
CHAPTER 6	69
Application of ET to a cosmic central body system	69
6.1 Planetary and galactic systems	69
6.2 The field solutions of the three manifolds of the central body.	70
6.2.1 Field solution of the outermost manifold of the central body	70
6.2.2 Field solution of the innermost manifold of the central body	70
6.2.3 Field solution of the central body manifold	70
6.3 Circular motion of a satellite of mass, negligible in comparison to the central body mass	71
6.4 Use of equation (37) in §6.3 to obtain the contribution made by the innermost manifold to the rotational speeds of stars in the M33 galaxy	73
6.5. Gravitational lensing	73
Discussion and Conclusions	75
References	97
Bibliography	99

PREFACE

I am Deepal Benaragama and I live with my wife Lalitha in the county of Kent, England.

I have a Ph.D. in applied Electromagnetism from the University of Southampton, England. Since leaving the university, I have worked for G.E.C. in the Midlands and C.E.G.B and Nuclear Electric in South East England. I took voluntary severance from Nuclear Electric, my last place of work, in 1993 when the research laboratory moved to a faraway location to which I was not prepared to go for personal reasons.

Voluntary severance allowed me to work full time in theoretical physics. Until then my work in this subject had been limited to a critical survey of the existing structures of Relativity and Quantum Mechanics, and during that survey, the outline of a complete mathematical theory of the physical world began to appear in my mind. In 1999 I wrote down these emerging ideas in a book I called 'The Quest for Balance' (An Extended Theory of Relativity). It was really an attempt to streamline my ideas and I self-published it using my own publishing company Kyral Ltd. It was never intended for sale and I distributed it freely among some of the professional physicists, my friends and colleagues. I received only one response and it was from a distinguished professor of physics who advised me to write a peer-reviewed paper based on my ideas. I am very grateful to him.

The complete mathematical theory of the physical world this book contains, which I call the Eigen Theory, is the results of my continued efforts over the last two decades. It finally turned out to be an intuitively appealing theory, which calls only for a basic knowledge of relativity and differential geometry.

After the Introduction, the reader may proceed directly to Chapter 4 in which the theory formally begins.

D S Benaragama
March 2020

ACKNOWLEDGEMENTS

Throughout this project, I have had support of friends, colleagues and relatives, and I owe them a deep debt of gratitude.

I thank Dr. Edvige Corbelli (ref.3) for promptly emailing to me the set of M33 data that I used for obtaining the curves in Fig .2.

Without the generous encouragement and support of my wife Lalitha, this book, which took more than two decades to complete, would not have been a possibility.

Finally I thank the staff at Grosvenor House Publishing, especially the assistant manager, Becky Banning, for her professional help and her kind, patient acceptance of the numerous amendments I made to the text.

CHAPTER 1

Introduction

According to Special Relativity (SR) which unified Newton's space and time the speed of light is an absolute constant in all inertial frames. Hence, the spacetime structure of SR is such that the relative motion which is characteristic of inertial frames and the absolute motion of light form a pair of opposites. This structure of SR opened up the possibility that other similar pairs of physical opposites could also exist. According to Niels Bohr such opposites are not limited to just the physical world. For he chose Contraria sunt complementa (opposites are complementary) as the motto for his coat of arms. The following quote, which this book fully substantiates, indicates the depth of Bohr's perception of opposites.

There are trivial truths and great truths.
The opposite of a trivial truth is plainly false.
The opposite of a great truth is also true. -Niels Bohr.

In the Eigen Theory of the physical world (ET) that this book presents any pair of great, or fundamental, opposites are complementary, or to be precise they balance each other. In the simplest form of this balance each of the pair of opposites acts as the fulcrum on which two states of the other balance. ET begins with the simplest of these pairs of opposites which is point and line in the Hausdorff space of *n*-dimensional spacetime $\mathbf{R}^n$ (ref. 5).

These opposites point and line are:

(a) the Occam's razor of the geometry of *n*-dimensional spacetime $\mathbf{R}^n$

(b) the symbolic geometrical representations of rotation and translation

(c) the geometrical counterparts of quantum and continuum, or particle and field, which are characteristic of Quantum Mechanics (QM) and General Relativity (GR) respectively. In Chapter 4 balance between point and line develops into that between particle and field. As a result, ET restructures QM and completes GR.

(d) related to the special elements of the two fundamental groups of addition and multiplication, 0 and 1, as these are the dimensions of point and line.

1.1 The state of balance between point and line

According to ET the state of balance between point and line is such that each of these acts as the fulcrum on which two states of the other balance. In Chapter 4 this balance develops into a pair of eigen velocity vectors (v, u) that transport each other producing the wave aspect of the wave-particle duality of matter. For this mutual transportation the energy of translation and a scalar angular momentum of rotation that the pair of eigen velocities mutually possess add up to zero in perpetuity. This is the only conservation principle in ET. It implies that translation (T) and rotation (R) of (v, u) are in a state of unification.

In standard physics two principles of conservation exist. According to these energy and angular momentum exist as separate entities of unknown fixed magnitudes. These are simply due to the present state of an unconditional separation of translation and rotation, or an unconditional T-R duality which goes counter to the condition of nothingness that may exist at the beginning of the physical world. As a matter of interest, this unconditional T-R duality caused much confusion to both Newton and Einstein and they were unable to resolve it to their satisfaction.

Each of v and u acts as the reference for the other which means in 2-dimensional spacetime (v, u) are Lorentzian boosts of each other (Appendix A). Hence according to ET, the concept of inertial frame central to the relativity of motion in SR is purely hypothetical and so is the absolute velocity of light *as a feature that accompanies it*. However, the invalidity of these hypotheses does not invalidate GR as it fundamentally applies only to empty spacetime. Also, the principle of relativity of motion that Galileo so succinctly put in his thought experiments known as Galileo's ship remains valid if what it actually implies is the presence of a built-in reference for motion just as in the case of the pair of eigen velocities (*v*, *u*).

The pair of velocities (*v*, *u*) is a function of a pair of tensors (*g*, *h*) where *g* is a pseudo-Riemannian symmetric metric tensor and *h* is an antisymmetric tensor. The latter is the sum of a 2-form and the exterior derivative of a 1-form. This 1-form is the counterpart of the set of arbitrary elements of the metric tensor which exists owing to the Bianchi identities that the metric tensor satisfies. The tensors (*g*, *h*) which stand on an equal footing are the fundamental elements of the gravitational and electromagnetic fields that constitute the internal structure of matter. The *g* field is the fundamental structure of the gross element of mass, the 2-form field is that of the gross element of charge and the field of the exterior derivative of the 1-form is that of the associated electromagnetic field. The set of field equations that (*g*, *h*) satisfy produce a unique solution of (*g*, *h*) only if spacetime is 4-dimenional. In standard physics 4 dimensionality of spacetime is simply assumed.

While (*v*, *u*) possess T-R unity as described above, (*g*, *h*) possess T-R duality as (*g*, *h*) moves with a pair of translational and rotational velocities which are distinct and of magnitudes 1 and 0, respectively. The T-R unity of (*v*, *u*) is the fulcrum on which this T-R duality of (*g*, *h*) balances.

As a point of interest, the present forms of gravitational and electromagnetic fields are those present in empty spacetime that just surrounds matter.

1.2 The fundamental structure of the physical world as the state of balance between particle and field

The mutual transportation of (v, u) for which each velocity acts as the reference for the other is the primary mode of motion in the physical world and it takes place in a state of nothingness. Therefore, the pair of eigen velocities (v, u) appears to exist by itself and creates its own spacetime manifold by mutual transport. Nonetheless (v, u) is not complete by itself as its state of balance is not complete. For, as stated earlier, presently this balance is such that, the T-R unity of the wave aspect of the wave-particle duality of matter is the fulcrum on which this T-R duality of field balances. For balance to become complete a T-R unity of this field should also act as the fulcrum on which the T-R duality of the particle aspect of the wave-particle duality of matter balances.

The wave aspect of the wave-particle duality matter that the pair of eigen velocity vectors (v, u) produce is already poised for the occurrence of the second balance mentioned above. For the pair of eigen velocity vectors (v, u) has an implicit field of a pair of tensors (g, $\underline{h}$) the first of which is a pseudo-Riemannian symmetric metric tensor and the second, an antisymmetric tensor. Just as in the case of h, $\underline{h}$ is also the sum of a 2-form and the exterior derivative of a 1-form. This tensor field ($\underline{g}$, $\underline{h}$) which is implicit for the wave aspect of the wave-particle duality of matter becomes explicit and characteristic of a base manifold without which the particle aspect of the wave-particle duality of matter cannot exist.

The base manifold consists of two parts which are termed, the innermost and the outermost manifolds of the particle aspect of matter. The antisymmetric tensor field of the innermost manifold consists only of the 2-form field; hence the innermost manifold may be classed as dark owing to the absence of the exterior derivative of the 1-form field. This field structure of the innermost manifold is the ET equivalent of the structure of dark matter. The outermost manifold, the antisymmetric tensor field of which consists only of the exterior derivative of an arbitrary 1-form field, is the same as the empty spacetime of GR.

The particle aspect is a splitting of the wave aspect of matter into a pair of alternative matter and antimatter particles which alternatively share the manifold characterised by (g, h). However, with respect to the base manifold the two particles have their own manifolds with their own tensor fields (g, h) and become concurrently present at their own separate locations in the base manifold. These two particles are the progenitors of the proton and the electron of the Hydrogen atom.

Each particle manifold is made up of the following three pairs each of which is a continuous distribution within the particle manifold.

time – space
translation - rotation
pseudo-Riemannian symmetric tensor g - antisymmetric tensor h

Each pair of translational and rotational velocities is orthogonal with respect to the symmetric metric tensor g.

Therefore, each particle possesses T-R duality. In other words, while the wave aspect of the wave-particle duality of matter possesses T-R unity, the particle aspect possesses T-R duality. A particular pair of the translational and rotational velocities of the tensor field (g, h) of the particle manifold unifies at the particle centre. This particle centre acts as the fulcrum on which the distribution of the pairs of translational and rotational velocities of each particle balances. This is a case of a T-R unity of field acting as the fulcrum on which the T-R duality of matter balances.

As a fulcrum of balance is primarily independent of those that it balances, particle centre primarily occupies only the base manifold, to be precise it occupies the innermost manifold. Hence the innermost manifold is shared by the two particle centres. The outermost manifold which is the same as the empty spacetime of GR externally envelops both particle manifolds. The two particles via this sharing of the innermost and the outermost manifolds exist in a state balance. As mentored earlier, these two particles are the progenitors of the proton and electron of the Hydrogen atom.

The occupation of the base manifold by the two particles manifolds is such that the two pairs of base manifold tensors ($\underline{g}$, $\underline{h}$) present at the two locations of the base manifold occupied by the two particle centres become respectively the same as the two pairs of tensors (g, h) present at the two particle centres.

The sharing of the base manifold by the two particles places severe restrictions on both the base manifold and the two particle manifolds. Hence the reason that the proton in the Hydrogen atom behaves as a central body and the electron behaves as its satellite. In this case the base manifold, which has become the monopoly of the proton, is comparatively unaffected by the presence of the electron.

From the foregoing it follows that the particle centre is of an extraordinary character. In standard physics, where particle is just a point, this extraordinary particle centre becomes overwritten by this simple point. This draconian measure is the primary cause of the present distorted view of the physical world that standard physics represents.

1.3 A simple view of the field structure of a particle of matter

Putting it quite simply, the combined gravitational and electromagnetic field structure of a particle is such that the *g* field in its simplest form is a spatially spherical container that holds the 2-form field captive inside it. The force of natural radial contraction of the *g* field would then be balanced by the force of natural radial expansion of the 2-form field. The exterior derivative of the 1-form field would extend beyond the spherical container, or particle of matter, into the empty spacetime that surrounds it and merge with its gravitational field which satisfies the empty spacetime field equation in GR that allows for the existence of an exterior derivative of an arbitrary 1-form field.

According to the above, one gets the impression that *g* and *h* fields of the particle manifold are in a state of balance that pervades the entire physical world from atoms to galaxies. However, the balance in this case is complex as it involves a base manifold consisting of an outermost manifold and an innermost manifold that the particle manifold occupies. The gravity field of the outermost manifold is the same as the empty spacetime gravitational field of GR; hence it decreases with increasing distance from the particle centre. The gravity field of the innermost manifold is the very opposite as it increases with increasing distance from the particle centre.

The reader may now directly proceed to Chapter 4 where the Eigen Theory formally begins.

CHAPTER 2

Balance

2.1 Symmetry misinterpreted as balance

Symmetry is often thought of as a precise statement of balance. However, that balance is only static whereas true balance is dynamic also. The fulcrum associated with this balance is the counterpart of the one in the old-fashioned merchant's balance; in both these cases the fulcrum-body, not the fulcrum, has no limits placed on its shape and size and makes no direct contact with those that it balances. Thus, the fulcrum body is a singular aspect of balance that could all too easily escape notice.

The concept of balance in the physical world first occurred in the law of equality of action and reaction, which was Newton's own contribution to the general framework of physics. This law belongs to the category of things in life that are well-nigh folklore, but also the least thought of in-depth. Implicit in it is the idea that fundamentally, things, which in this case are bodies and forces, never occur singly, but as pairs that interact not chaotically but with the precision of nullity. Apparently, the pursuit of excellence in physics is also the pursuit of this balance in the physical world, as evinced by the change over from Newton's physics to Einstein's relativity. For this change over, the balance changed from that of two equal and opposite forces to that of a null ray of light which acts as the fulcrum on which space and time balance.

2.2 Balance as the highest Principle of nature

According to this book, no principle exists which is higher than the principle of balance; hence it is the most elusive to pin down. Because of this adversity that balance presents, it appears that the human mind with its power of abstraction and sense of formal beauty has taken charge and has established polarisations such as symmetry and antisymmetry as truth. However, the human mind, in spite of its supremacy and power over most things, is still just a component of life. Therefore, what appeals to it, even in the guise of a fundamental necessity for itself, needs careful inspection before it could serve as a basic feature of the physical world that possesses a clear measure of objectivity. In fact there are good all round reasons which suggest that polarity-appeal in itself is a balancing act being performed at the deepest inner recesses of what may be called the 'human psyche' for want of a better phrase. Ever in motion and in change, the restless mind yearns to embrace that which is immovable and invariant in its relentless struggle to reach a state of balance. So it seems that the rules of balance by which even mind has to function are accessible only from a stratum of intuition which lies above both mind and matter and controls the behaviour of both these.

According to the above, even those high ideals such as morality and liberty that we hold very dear to us see balance as their mentor; this balance is what becomes developed in this book. For this development, the cue comes from how balance operates in the physical world, and to establish this operation use is made of every morsel of firmly established knowledge in physics. This book makes it clear that physical balance is able to bridge the outer reaches of the cosmos and the inner recesses of the atom and complete the program that Newton initiated with what was available to him in his time. In simpler, colourful words, it would become clear that balance drives both galactic systems and heat engines; it tenders equally to both rust and rose petals.

Balance is certainly not stasis; clearly, it is also different from total and perpetual change. Stasis and change find common ground in balance. Such complexity of balance is what instils the feeling of chaos in some, and indeed, the opposite feeling of order in others. The feature that contributes mostly to these extreme views is the singular nature of the fulcrum body involved in the process of balance.

As mentioned earlier, the fulcrum body is a truly singular feature in that it can be as large as the Earth itself at base, but must reduce virtually to zero at top and turn into the fulcrum. In that sense fulcrum body is, as the saying goes, both Alpha and Omega as far as those that it balances are concerned. There is such a fulcrum body for the ultimate physical-balance, and there is also a fulcrum body for the ultimate life-balance and it accounts for the apparently singular aspect of life that often throws regular life into confusion. The cause of much of the controversy about life, be it science against religion, theism vs. atheism, to mention just two, is the failure to comprehend and come to terms with the balanced state of life.

Because of life's apparent singularity, there is always more to life than even the most profound of our systems of beliefs, thoughts, or even empirical and theoretical discoveries. Even so, they are also the stepping-stones, which allow us to reach out to and realise life's ultimate state of balance, which amounts to *'doing the right thing according to the prevailing circumstances no matter how difficult that action turns out to be'*. This deceptively simple dictum can also be very difficult to put into practice, especially when it matters the most. In such circumstances, access to a clear vision of the blue print of life is what we require. Most would agree that in our age of knowledge and reason, acquisition of this vision could, and should, begin with physics. Accordingly, in this book a *sui generis* theory of the physical world develops which subsumes relativity theory in somewhat the same way that the latter subsumes Newton's theory. It clearly shows that Quantum Mechanics despite its remarkable strength and accuracy is still mainly a tool, which has been developed to cope with the present incomplete knowledge of the basic architecture of the physical world and the failure as yet to capture its latent beauty and 'simplicity'.

This book also aims to dispel an asphyxiating myth that has gripped physics by its throat. This myth is that 'reality' is so different and distant from ordinary living that it can only be the territory of privileged classes such as academics, and conglomerates operating super expensive experimental and observational rigs. On the contrary, 'reality' ignores educational, social or economic distinctions. It makes no distinction between the basic architecture of a particle of matter and that of a living being. In our past rational awakening, Newton saw this very clearly, when he said in his 'Opticks' that 'Nature is very consonant and conformable to herself'. Often it requires a great deal of patience to sit and think for as long as it takes to notice this remarkable conformity in life. Just to take a simple but basic example, both a particle of matter and a living being would optimise the path that they would traverse to go from a location A to a location B. On an even more fundamental note, the physical world is a 'love affair', or a pair formation, between proton and electron just as it is also with the animal world of male and female. Proton purveys the element of stability that is essential for the perpetual dance of the electron just as it is also in the animal world where the male is often the stabilising element, and the female, the element of activity, colour and chemistry. There are many more tantalising parallels. Every fundamental feature of the sub-atomic world or of the vast reaches of the cosmos has its reflection in our ordinary lives and vice versa, and to trace these parallels is an exhilarating adventure that no rational thinking human being should miss.

Adventure can lead to discovery, and so it is in this case. Existence of fundamental parallels such as the ones mentioned above signals that the deepest fundamental features of the theoretical basis of the physical world might have life-parallels that would make it easy to see the right or the balanced path that we should follow in life. For example, the fundamental theoretical basis of physics is strictly local, and is free of global speculations. Instead of speculating on global features, allowing these to develop naturally in terms of those that are strictly local has been the proven procedure. Euclid's fifth postulate is a good example of a global speculation that may even have held back geometry for quite a while. Therefore, it seems that the rational basis on which we formulate our path in life should be strictly local. A stance which opens up global questions such as 'whence come I and whither go I' is to be abandoned in favour of one that queries, 'Now that I am here what should I do'. The next conscious and balanced step that we take is what matters, and if there were such a thing as a universal consciousness, then it would respond autonomously. There is so much to achieve in the here and now to construct a balanced frame of mind and no time to waste, so the spirit of science whispers in our ears.

Often it is the scientist, on account of being forced to focus on the exception rather than the norm in an effort to bring clarity to the latter, who misses out on this adventure and therefore it is they who sometimes have to, in the end, struggle harder to regain their balance. Perhaps owing to the brutal nature of this struggle, contemporary physics does not portray a world that bears much resemblance to the real world of intuition.

For example what physics, even if it is computationally accurate in a certain specific area, has any luminous reality if it implies that the moon does not exist if it is not looked at and that a single object can be at two different places at the same time? Even that computational accuracy has come at a great price, the price of contriving to cancel out point-particle infinities by creating more infinities, or the price of carelessly sweeping infinities under the carpet, as it were, in the case of string and superstring theories. For in these theories the string can only ultimately come into being as the result of a point moving with infinite speed, if it is, indeed, assumed that there can be no other premise which is more basic to physics than the displacement of an event in spacetime.

Just as fundamental particles of matter have eventually formed into a cosmos, so have fundamental ideas of physics now evolved into an extreme state of a cosmos of its own which is as far removed from the everyday human thought as is the cosmos of stars and galaxies from ordinary matter. Yet it produces remarkable results, or facets of truth, without which it would be impossible to formulate the basic layout of the physical world: rather like, it would be impossible to 'formulate' life without the higher elements that are being cooked in the raging infernos at the cosmic end of the physical world. The present ***distant state of physics*** is mainly due to symbolic abstraction that is characteristic of the world of mathematics. It is as if literature has had to step aside to give way to complex and incoherent fabrications. Here in this book literature has the final word.

CHAPTER 3

A Brief History of Physics

3.1 The early stages of physics

Physics is perhaps as old as the humanity itself and it too has gone through a number of recognisable evolutionary stages. The first of these, which distinguished physics from other human arts and crafts, may have occurred around 3000 BC in astronomical observations carried out in Egypt, Mesopotamia, India and China. These observations also appear to be the first recognisable scientific activity of the human awareness of nature. From those fragments of ancient history available to us, it is possible to guess what really led to it. Firstly, there were good practical reasons that were helpful to the human survival and prosperity; timekeeping for activities such as farming and making astrological predictions appear to have ranked foremost among these. Secondly, faced with the harsh reality of nature the ancients may have found solace in the thought that astronomical observations enabled them to commune with the heavenly activities and thereby remain just that much closer to the home of the gods that sparkled in the sky night after night.

Heavenly astronomy had an unsuspecting earthly companion, which was geometry or the art of measuring the earth. Practical geometry began in ancient Egypt as rope-stretchers annually re-established the boundaries of the farmlands that were flooded by the River Nile and it gradually became a refined tool in the construction of the Egyptian pyramids. However, geometry remained a mere practical tool until the Greek civilisation, which appeared around 800 BC, distilled it into an abstract form. The process of deductive reasoning, which promotes a healthy curiosity of the knowable in place of a morbid fear of the unknowable, is the principal character of the next stage of physics.

Around 600 BC it appears that the human perception of the universe began to change radically, profoundly and decisively. The epicentres of this change were in India, China and Greece. Around this period, these countries produced men who saw the universe not as a chaotic playground of numerous capricious gods, but as an orderly cosmos, which functioned harmoniously with all living things. In India and China, the vision was sudden and metaphysical, but in Greece, the new vision of the universe developed much more gradually, involved generations of thinkers, and had the strong undertones of a study aimed at a detailed exploration of the material aspect of the universe. In laying the foundation of this study, all they had was their imagination, and they unleashed it in a torrent of magnificent ideas, some of which continue to hold their ground to this day.

The gradual process of philosophic change in Greece began with the revolutionary thoughts of a Milesian intellectual giant called Thales (c.624-546 BC). Thales primary aim was to replace the supernatural with the natural through observation and experiment, but his lasting contribution was converting geometry from an art of measurement into an abstract science of deduction based on general propositions; those given below are attributed to him.

(a) An isosceles triangle has two similar angles at its base. (The word similar suggests that Thales may not have been able to assign a magnitude to angle)

(b) The opposite angles at the intersection of two straight lines are similar.

(c) The base and the two angles at its ends determine a unique triangle.

(d) A circle is divided into two equal parts by a diameter.

(e) A diameter of a circle subtends a right angle at any point on the circumference.

Although such knowledge of geometry is rudimentary and obvious to us now, it must have been an enormous feat of imagination at that time.

The next Greek thinker who made a substantial contribution to physics was Pythagoras (c.570-500 BC), from Ionia. He and his followers made a truly fundamental discovery in geometry, which has become known as the 'Theorem of Pythagoras'. It says that 'the square on the hypotenuse of a right-angled triangle is equal to the sum of the squares on the other two sides'. Pythagoreans were said to be the first to have imagined that the earth and all other heavenly bodies had spherical shapes. They also believed that the earth, the sun and the other planets all moved round a 'central fire' and that these movements were accompanied by an inaudible music, which they called the 'harmony of the spheres'. They taught that light consisted of particles, which travelled from an object to the eye.

At the end of one and half centuries of existence, the Pythagorean School started to dwindle and around 380 BC a new Athenian school, called the 'Academy' arose. The founder of this school was the Athenian philosopher Plato (429-347 BC) in whose fertile mind mathematics, and in particular geometry, soared to the heights of divine knowledge. To Plato and his followers, abstract geometrical forms became the true substance of the universe, and the 'Academy', which lasted for nearly a thousand years, boasted above its entrance door the motto 'Let none but mathematicians enter here'. Naturally, geometers fashioned the heavens using abstract geometrical forms, and astronomy and geometry finally shook hands.

By the middle of the fourth century BC, science in Greece had begun to decline, and, in about 300 BC, the centre of scientific excellence moved from the Academy, in Athens, to the Temple of the Muses, the equivalent of a modern university, in the city of Alexandria, Egypt. Here Euclid (c.330-275BC) produced his 'Elements of Geometry', which consisted of a coherent treatise of thirteen books on geometrical 'truths' which may be regarded as the real foundation of the physical sciences. It was also here that the great mathematician, Archimedes (287-212 BC), had his education before returning to his native Sicily. Then there were the four great astronomers; Aristarchus of Samos (c.310-230 BC) who proposed a heliocentric solar system; Eratosthenes (c. 276-195 BC) the chief curator of the library of Alexandria, who was the first to measure the distance between the Sun and the Earth to a reasonable accuracy and was also said to have measured the 'obliquity of the ecliptic' or the tilting of the earth's axis of rotation which causes the seasons; Hipparchus (c. 190-120 BC) of Nicaea who discovered and estimated 'the precession of the equinoxes', or the minute wobbling of the earth's axis of rotation; and finally Claudius Ptolemy who produced an extensive treatise on mathematics and astronomy. This treatise, which became known as Almagest consists of a collection of thirteen books and describes a geocentric universe, which although completely erroneous, remained in authority until the sixteenth century AD. Science continued to flourish in Alexandria until the Dark ages began about 400 AD, when it became virtually dormant for a full thousand years.

Around 1400 AD concurrent with the literary renaissance in Europe, science awoke from its slumber and the third stage of evolution in physics began. Many regard Leonardo da Vinci born near Empoli (1452-1519), who was talented in virtually every field of human endeavour, as the first renaissance scientist; it appears that he studied nature in the same spirit as we do now. He had been an advocate of the view propounded by the Greek thinkers Democritus of Abdera (c.470- 400 BC) and Anaximander of Miletus (c. 611 -545 BC) that the universe is governed by mechanical laws. According to Leonardo, we should base science on observation, discuss using mathematics and verify using experiment. He was of the opinion that scientific certainty can only come from mathematical reasoning.

As mentioned above physics may have stirred to life through cosmic curiosity, but it certainly found its feet through cosmic order. The first detailed and systematic study of cosmic order was the geocentric system of the universe produced by Ptolemy. The Polish astronomer Nicolaus Copernicus, (1473-1543) sought to reduce the complexity of this Ptolemaic system, which consisted of about eighty circles of deferents and epicycles, by placing the sun at the centre of the system, rather than the earth. He worked very hard at his heliocentric model but it never occurred to him to question the prevailing belief in the 'naturalness' and 'inevitability' of the circular motion of planets. Because of this, his model too remained complex; but he managed to reduce the number of circles to thirty-four.

With hindsight we now know that all Copernicus had to do was to abandon circles in favour of ellipses; a minor adjustment which would have reduced the number of trajectories to seven and increased the accuracy of the model beyond all expectations. However, no amount of speculation could have bridged the gap that existed between the circle and the ellipse in astronomy at that time. Masses of accurate astronomical data, a matching mathematical acumen and a monumental perseverance were necessary for this purpose. The man who had the means and the inclination to produce such masses of data and satisfy the first of these three requirements was the Danish astronomer Tycho Brahe (1546-1601). He was weak in mathematics and strong in astronomical observations. To obtain better accuracy he used better equipment, repeated a measurement many times over, and obtained their average. Thus, he introduced a new standard of accuracy into astronomical measurements, and to this accuracy, he charted the positions of stars and planets for a period of over twenty years. However, the analysis of this data was beyond his ken and destiny placed it in the hands of Johannes Kepler (1571-1630), the brilliant German mathematician from Weil near Stuttgart. Only Kepler had the qualities, which satisfied our remaining two requirements.

In a tireless attempt to fit Tycho Brahe's observations on Mars to an analytic curve, Kepler shattered the groundless belief that heavenly bodies should, of necessity, move in circles merely because these display an elegant symmetry not found in other 'lesser' figures. He found that all the observations fitted perfectly not to a circle but to an ellipse, a curve that was believed to be imperfect. In 1609, he published his findings in a book called Astronomia Nova in the form of the following two laws, which applied to the orbit of Mars.

The planet moves in an ellipse with the sun at one of its foci.

The line joining the sun to the planet sweeps out equal areas in equal times.

In 1618, in a second book called Epitome Astronomiae Copernicae, Kepler next extended these laws to all the remaining planets, the moon, and the four satellites of Jupiter; in other words to all the planets and the satellites of the solar system known at that time. Again, in 1619 he published a third book, Harmonices Mundi, in which he announced the following third law, which too applied to all the planets.

The square of the time that any planet takes to describe its orbit completely is proportional to the cube of its average distance from the sun.

In replacing the circular orbits with elliptical orbits, Kepler answered the age-old question of 'how' the planets moved round the sun. However, in so doing, he naturally raised the question 'why' the orbits had to be elliptical and not any other geometrical curve. Of course, the circular orbits, owing to their naturalness, escaped this question completely. The best that Kepler could do was to imagine that whilst the sun provided the motive force for the planets to move round in circles, a force akin to magnetism pushed and pulled the circular orbits into elliptical shapes. More than half a century was to elapse before these qualitative conjectures were to become unnecessary in the face of a superlative theory put forward by Isaac Newton (1642-1727).

Whilst Kepler was discovering the architecture of the solar system, an Italian physicist and astronomer, Galileo Galilei (1564-1642), was investigating motion of bodies by conducting rudimentary terrestrial experiments and, where such experimentation was not possible, by resorting to ingenious thought experiments. Prior to Galileo's investigations it was believed that the natural state of a body was that of rest and that a force was necessary to keep it in motion. Galileo came to the firm conclusion that rest and uniform motion were indistinguishable from each other if an external reference were unavailable. He also concluded that a force just had the effect of changing these states by producing acceleration. Thus, Galileo was the first to conclude correctly, and demonstrate experimentally, the principle of relativity and the principle of inertia, which became corner stones of the early classical physics.

However, being unable to shake off the influence of a prevalent belief that certain motions, such as the fall of heavy bodies, result from a natural tendency that is a factor present in addition to inertia and force, Galileo was never able to state the principle of inertia with clarity. Nonetheless, Galileo was able to shatter the belief that gravity causes heavier objects to fall faster than lighter objects and to state quite clearly that all objects fell at the same rate. Until the time of Galileo, the analysis of the motion of a projectile, which formed an important part of the curriculum of mechanics, remained a daunting and an inconclusive task. Using his insight on inertia and the fall of objects under gravity Galileo was able to analyse this motion correctly.

Galileo pioneered the telescopic astronomy, using telescopes he made for himself. It is very likely that his were the first mortal eyes that were able to examine closely the surface irregularities of the moon, the stars which forms the Milky Way, the four satellites of Jupiter, the phases of Venus, and the rings of Saturn. However, he wrongly interpreted the rings of Saturn as two small spherical bodies touching Saturn at the opposite ends of a diameter.

3.2 Newtonian mechanics

Kepler and Galileo provided the key features of an emerging system of mechanics, of universal applicability, within which all primary physical distinctions between the heaven and the earth were to disappear. The formulation of this system of mechanics needed the insight of a genius of extraordinary talent. This genius was the physicist and mathematician Isaac Newton (1642-1727), from Lincolnshire, England. He reduced Kepler's three laws into a single universal law, set in a background of three completely general laws of motion, two of which embodied Galileo's principal discoveries. Newton described this universal system of mechanics in a book published in 1687 under the title 'Philosophiae Naturalis Principia Mathematica'; of this it can be justly said that, with the possible exception of Euclid's 'Elements' and the English naturalist Charles Darwin's (1809-1882) 'The origin of the Species', no greater scientific work has been produced by the human intellect, before or since. Newton's first law of motion was a succinct statement of the principle of inertia that Galileo almost had within his grasp. It was as follows.

Everybody continues in its state of rest or of uniform motion in a straight line, unless it is compelled to change that state by impressed forces.

Now a body may be at rest or be moving uniformly with respect to some location on the earth, but since the earth itself is moving non-uniformly with respect to the sun, the body has a completely different motion with respect to the sun. As this state of affairs could continue, it is clear that motion is not an easy concept to define. Newton was fully aware of this difficulty; as a way out, he defined an absolute inertial frame in terms of the following two postulates.

Absolute space, in its own nature, without relation to anything external, remains always similar and immovable.

Absolute, true, and mathematical time, of itself, and from its own nature, flows equably without relation to anything external.

Newton attributed absolute space to the presence of huge, immovable masses, which delimited the universe; a kind of immovable framework within which all movable bodies are to be found. Even Newton himself was ill at ease with this framework and with the absolute clock that ticked away regardless within it but, as there were no alternatives, they became established concepts and held their ground for a period of nearly two hundred years. Having made provisions for substantiating motion, Newton next went on to give a quantitative definition of force in the form of a second law of motion, as follows.

The rate of change of momentum of a body is proportional to the impressed force and is in the same direction as this force.

This law also introduced a new concept called momentum and Newton defined it as velocity multiplied by an invariant property of the body that he called mass. Thus, Newton introduced the fundamental property of mass to physics. Next, Newton introduced the following third law of motion, which completed the general description of force.

To every action there is always an equal and opposite reaction.

Unlike the first two laws that were in the making even before Newton appeared on the scene, this third law was of Newton's own making. This law is a statement of a precise balance that transcends time as the forces therein act instantaneously. As we shall discuss in due course, what Newton really saw was a transcendental principle of balance that operates timelessly in the physical world. Forces just happened to be the only medium of expression that was available to him in his time.

Newton's system of mechanics and the absolute reference frame, which he defined in terms of rigid measures of space and time, are still in use today as they had been for hundreds of years in the past. However, our present knowledge of the nature of space and time leaves us in no doubt that Newtonian mechanics is only valid for speeds that are very much smaller than the speed of light. At speeds that are significant in comparison with the speed of light, relativity theories take over. According to these theories, which were first formulated by Albert Einstein (1879-1955) in a special form in 1905 and then in a generalised form in 1916, motion is a relativistic concept which did not require the support of an absolute space and time: they, after all, are mere hypothetical concepts. However, a small trace of a similar hypothesis remains in relation to the rotary motion of bodies; this remnant is Newton's belief that the cause of the absolute nature of rotational motion lies in distant masses of the universe.

The single universal law, to which Newton reduced the three laws of Kepler, describes with a high degree of accuracy the universal phenomenon of gravitation. Newton's law for this phenomenon is simplicity itself, and it is as follows.

Everybody in the universe attracts every other body with a force proportional to the product of their masses and inversely proportional to the square of their distance of separation.

Newton's physics, which explained all the cosmic and terrestrial phenomena that were significant in his time, reveals four basic parameters, which can remain unaffected by varying physical conditions. These invariant elements of Newton's physical world are the mass, energy, momentum, and angular momentum of a particle of matter. The appearance of these four conservable physical parameters and the associated four laws of conservation mark the beginning of a scientific analysis of the intuitive notion of permanence, which we habitually associate with the physical world.

In addition to gravity, there were two other basic phenomena, electricity and magnetism, which had, since about 1600, drawn the attention of physicists. With the availability of Newton's framework of mechanics, and aided by his theory on gravity, electricity and magnetism began to develop as two separate sciences called electrostatics and magnetostatics. From 1821, the English experimental physicist Michael Faraday (1791-1867) began to conduct what we now call electromagnetic experiments, and after a period of about ten years discovered the link between electric and magnetic fields. In the fertile imagination of Faraday, the concept of physical field turned from fiction to fact. In 1856, the Scottish Physicist and mathematician James Clerk Maxwell (1831-1879) successfully arranged all the available knowledge on electromagnetic phenomena, together with a small but vital contribution of his own, into an elegant mathematical theory on electromagnetism, called Maxwell's Theory.

Meanwhile Newton's physics had been evolving and by 1834, it had developed into analytical mechanics. Many eminent people contributed to its growth, notably Leonard Euler (1707-1783), Joseph Louis Lagrange (1736-1813), William Rowen Hamilton (1805-1865) and Jacobi (1804-1851). A fundamental feature of analytical mechanics is that position and momentum replace position and velocity (in Newton's physics) as two independent base states. Thus, analytical mechanics gives momentum an identity of its own. Absence of any explicit forces is another important feature, which distinguishes analytical mechanics from Newton's physics. A third difference is that analytical mechanics is independent of coordinate systems. This feature enables a mechanical system to have coordinates of its own. These coordinates, which are the positions and momenta of the components of the mechanical system, define an abstract space called phase space in which the entire mechanical system appears as a single point. Finally, analytical mechanics is ideally suited for the analysis of complex mechanical systems whereas Newton's physics is not.

For a while it looked as if, within the theoretical framework provided by Newton's gravitation, Maxwell's electromagnetism, and analytical mechanics the physical world was about to unfold itself completely in terms of just two independent basic properties of matter: mass and charge. But that was not to be.

3.3 Relativistic physics

In 1887, Albert Michelson and Edward Morley established experimentally that the behaviour of Maxwell's electromagnetic waves was incompatible with Newton's absolute separation of space and time. In 1905, Albert Einstein (1879-1955) who may not have been aware of the Michelson-Morley experiment formulated a new theory of space-time that resolved this incompatibility and showed that Newton's view of separate space and time is flawed. Two others were individually active in this respect and they were the Dutch physicist Hendrick Antoon Lorentz (1853-1928) and the French mathematician Henri Poincare (1854-1912). In 1908, the Russo-German mathematician Hermann Minkowski (1864-1909) found the correct interpretation of Einstein's theory and unlocked its full potential. According to Minkowski interpretation, space and time are not separate entities as Newton had imagined, but conditional partitioning of one continuum, which we now refer to as spacetime. With respect to this new continuum, the concept of position, which used to consist of just three space-coordinates, becomes broadened to include the time coordinate on an equal footing. This wider concept of position is termed 4-position or, simply, event. Similarly, velocity and momentum, each of which used to consist of three components, acquire a fourth component and become 4-velocity and 4-momentum, respectively. A technically more correct name for the latter is momentum 1-form. The fourth component of 4-velocity prima facie carries no special significance, but that of momentum 1-form represents the energy of matter. Thus, the energy and momentum of a particle of matter, which are conserved as two separate entities in Newton's physics, become a single conserved entity in Einstein's theory, which became known as special relativity. Furthermore, the magnitude of this single entity becomes the energy equivalent of the rest mass of a particle of matter. Thus, in special relativity, three of the four conserved entities in Newton's physics unify, but the fourth entity of angular momentum remains separate.

The fundamental changes that position, velocity, and momentum undergo in special relativity are concomitant with the speed of light becoming an absolute entity and occupying a special position in physics as required by Maxwell's theory. More precisely, special relativity is an exposition of the spacetime structure that Maxwell's theory demands and they fit like hand and glove. For, in principle, Maxwell's theory is equivalent to special relativity supplemented by just the Coulomb potential of charge.

In 1916, Einstein's thoughts on the nature of gravity led him to a generalisation of special relativity, called general relativity.

This generalisation clearly showed that the rigid, flat space-time of special relativity is just the limiting form of a small, local region of a flexible, globally curved space-time. Einstein with his general relativity brought to an end physicists' belief in the supremacy of Euclidean geometry and the uniqueness of the Euclidean straight line, just as Kepler with his 'humble' ellipse brought to an end astronomers' belief in the supremacy of the circle and the perfection of the circular shape. A body, which freewheels along a 'straight' line in the curved spacetime of general relativity, would produce the apparent impression that it is under the influence of an invisible force. Thus, just as Newton's force of gravity explained Kepler's elliptical orbits of planets, Einstein's curved space-time explained Newton's concept of force. With this immense achievement, Einstein showed that the entity that we call the physical field consists of a component that we cannot construct accurately using Newton's concept of force.

General relativity established that the structure of this field component, which forms at least the major part of the large-scale structure of spacetime, originates in an entity called a metric tensor which is present as a continuous distribution (or a field) throughout spacetime. A metric tensor, in general, consists of a set of sixteen numbers. However, a certain property of symmetry that the metric tensor possesses effectively reduces this set to ten numbers. Thus, in relativity theory, an event, which carries a label of four arbitrary coordinates, receives ten more numbers, which represent what may be called its metric infrastructure. In a curved spacetime the metric tensor varies from event to event no matter what coordinates are used; in special relativity, it remains effectively constant throughout spacetime.

A spacetime endowed with a metric tensor field is termed a metric manifold. The term 'metric' signifies that this manifold is able to produce an objective measure for length that we can transport unchanged from one event to another. Also, in a metric manifold a unique line of optimal length, called a geodesic, exists locally between two arbitrary events. A geodesic is the curvilinear equivalent of the familiar straight line in Euclidean geometry. Since in Euclidean space free-motion describes straight lines, it follows that in a metric manifold it describes geodesics. The motion along a geodesic is expressible using either the velocity or the momentum, since we can change these from one to the other using the metric tensor without which the geodesic has no real existence. Thus, in general relativity there is no substantial distinction between momentum and velocity.

At present, general relativity is not fundamentally equipped to assimilate matter into spacetime, at least not in the way Einstein had wanted it, according to his book, 'Out of My Later Years'. Only a conjectural equation that Einstein himself proposed determines the metric tensor field, but only in empty spacetime. Consequently, general relativity is still incomplete and Einstein unreservedly maintained to the end of his life, that its completion entails just uniting gravitation and electromagnetism in a single structure. Einstein believed that this structure would be the Theory of Everything that happens in the physical world. The Eigen Theory completely vindicates Einstein's belief. However, it preserves General Relativity as it applies to only empty spacetime.

3.4 Quantum mechanics

According to the Eigen Theory, the physical universe is the embodiment of a 'unique' state of integration between generalised forms of gravitation and electromagnetism, respectively.

In contemporary physics, gravitation and electro-magnetism form two separate rival schools of thought known as General Relativity and Quantum Mechanics. At present Quantum Mechanics is in vogue and its adherents advocate subsumption of General Relativity into its body; the result would be a particle centred physical world, which would multiply complexities such as those involved in quantum entanglement. Simple, compelling mathematical reasoning exist, for instance Hausdorff Space, which suggests that continuation and quantisation, which are the substrata of gravitation and electromagnetism respectively, exist as a syzygy, or an archetypal pairing, and that this syzygy can only be reached from its continuity side and not from its quantum side.

Two crucial differences exist between Quantum Mechanics and the two theories of Newton and Einstein. Firstly, Quantum Mechanics leaves us with a feeling of utter confusion, at best with emptiness, whereas the two theories of Newton and Einstein appeals to our intuition and also they have a sense of uniqueness in that they rise or fall by their core structures within their areas of applicability. Tinkering with the core structures is not an option. Secondly, whilst Quantum Mechanics has always been a collective endeavour, each of the two theories of Newton and Einstein has been effectively a personal vision. Perhaps this may mean that in any epoch Truth reveals itself only to one for the benefit of the many. That said, we should leave alone the chaotic and vibrant world of Quantum Mechanics, and should access its vast superlative knowledge base only strategically and at opportune moments.

CHAPTER 4

The Eigen Theory of the Physical World

4.1 Introduction

At present two separate theories of physics exist, namely Quantum Mechanics (QM) and General Relativity (GR), centred on the concepts of quantum and continuum respectively. As these concepts are opposites QM and GR have remained separate in spite of ongoing efforts to bring them together as the complete theory of the physical world. However, according to the Eigen Theory (ET) this book presents opposites do come together in a state of balance and so do quantum and continuum beginning with their respective geometrical counterparts of point and line.

Point and line are the Occam's razor of the geometry of n-dimensional spacetime $\mathbf{R}^n$. In ET each of these is the fulcrum on which two states of the other balance, which is the simplest form of balance between a pair of opposites. ET begins by applying this principle of balance to the following description of point and line of the Hausdorff space given in Ref. 5.

The idea that the line joining any two points of $\mathbf{R}^n$ *can be infinitely subdivided* (*which means continuous*) *can be made* more precise by saying that any two points of R^n have neighbourhoods which do not intersect.

In ET the above two points of R^n joined by an optimum line develops into the foundation of the physical world in which QM becomes restructured and GR becomes complete. This development takes place in three stages in **§4.3, §4.4** and **§4.5**. A comprehensive introduction to these three stages is given below in **§4.2**.

4.2 A comprehensive introduction to the three stages of the Eigen Theory

4.2.1 Wave aspect of the wave-particle duality of matter

Let two points **P** and P_1 of *n*-dimensional spacetime R^n be joined with an optimum line of length *s*. Since the neighbourhoods of **P** and P_1 do not intersect, if **P** and P_1 tend to coincide then the line tends to acquire a limiting length *ds* which is not an increment of *s*. Accordingly, *ds* is the fulcrum on which the two points **P** and P_1 balance; *a case of balance between point and line in which line acts as the fulcrum on which two states of point balance.*

According to the principle of balance stated in **§4.1**, *a case of balance between point and line also exists in which point acts as the fulcrum on which two states of line balance.* In this case the velocity *d/ds* that corresponds to *ds*, say *v*, exists in a state of balance with a second velocity *d'/ds* that also corresponds to *ds*, say *u*, that balances **P** with another point P_2. In this case the point **P** is the fulcrum on which *v* and *u* balance. These are a pair of eigen velocity vectors that reference each other. Appendix A formulates the following set of equations F_1 that they satisfy.

$$\beta g_{ij} v^j = (g_{ij} + h_{ij}) u^j \qquad (A14)$$

$$\beta g_{ij} u^j = (g_{ij} - h_{ij}) v^j \qquad (A15)$$

F_1 produces n pairs of (v, u) and the corresponding n eigen values β each of which is (a) the magnitude of each of v and u (b) the hyperbolic secant of the angular separation between v and u. Also g and h are a pseudo-Riemannian symmetric metric tensor and an antisymmetric tensor respectively. The tensor h is the sum of a 2-form and the exterior derivative of a 1-form which is the counterpart of the n arbitrary elements of g which exist due to the Bianchi identities that g satisfies; hence g and h have the same number of elements and stand on an equal footing. According to F_1 in Minkowski spacetime v and u are Lorentzian boosts of each other. Hence the principle of balance even includes special relativity.

Tensors g and h are characteristic of gravitation and electromagnetism respectively. More precisely g and the 2-form of h are the respective field elements of the gross properties of matter, mass and charge and the 1-form of h is the ET equivalent of the Maxwellian electromagnetic potentials.

$|\beta|$ is confined to the closed interval [0, 1] in three ways (*a*), (**b**) and (**c**) as follows.

(*a*) $|\beta|$ within the open interval (0,1).

In this case (v, u) transport each other across spacetime producing a dual congruence of worldlines This dual congruence is the wave aspect of the wave-particle duality of matter.

F₁ in §4.3.2 produces the following

$$(g_{ij} + \check{h}_{ij})u^j v^i = 0 \tag{2}$$

$$\check{h} = h/(1 - \beta^2), \quad |\beta| \neq 1 \tag{2a}$$

Accordingly, energy and a scalar angular momentum of (*v*, *u*) add up to zero in perpetuity. This is the only conservation principle in ET; it implies that translation and rotation (T and R) of the motion of (*v*, *u*) unify in a state of nullity.

According to the nature of the points of *n*-dimensional spacetime $\mathbf{R}^n$ outlined in §4.1 the point **P** at which *v* and *u* intersect possesses a neighbourhood within which it is free to move arbitrarily. Therefore, **P** can be infinitesimally perturbed arbitrarily without disturbing the T-R unity of (*v*, *u*). In Appendix B this condition, *which is an expression of balance between point and line in which a point acts as the fulcrum on which two states of line balance,* produces the transport equations that (*v*, *u*) satisfy.

(b) $|\beta| = 1$.

In §4.3.3 the above transport equations of (*v*, *u*) transform into the following field equation **F₂** that produces (*g*, *h*) at **P**.

$$\{R^k_{ijn} + C^k_{ijn}\}\overline{w}^i = 0 \tag{11}$$

This transformation entails *v* and *u* coinciding as a field velocity vector $\overline{w}$ at **P**. Correspondingly $|\beta| = 1$ and $\overline{w}$ is of unit magnitude. (**R**, **C**) are functions of (*g*, *h*) and these remain invariant for a reversal of the sign of *h*. Hence $\overline{w}$ is the translational velocity of the field (*g*, *h*) at **P**. **F₂** restricts spacetime to 4 dimensions; hence 4-dimensional spacetime is a natural feature of ET. Contraction of $\mathbf{F}_2$ with respect to k and n or k and j results in the following equation which produces $\overline{w}$.

$$\{R_{ij}+C_{ij}\}\overline{w}^i = 0, \quad \det(R+C) = 0 \tag{14}$$

(b) $\boldsymbol{\beta} = 0$

For $\boldsymbol{\beta} = 0$, $\mathbf{F}_1$ becomes the following equation $\mathbf{F}_3$

$$\{g_{ij} \pm h_{ij}\}\overline{w}^i = 0, \quad \det(g \pm h) = 0 \tag{14a}$$

where w is an alternative pair of null rotational velocity vectors present at **P**. In ET the null magnitude of w, which is associated with g, simply has the effect that the spatial dimension of spacetime turns into the temporal dimension and not that these two dimensions of spacetime become unified into a 'transcendental' or 'mysterious' dimension as commonly thought at present. With this interpretation of w spacetime becomes free of null geodesics and also the accompanying unnatural 'time-like' and 'elsewhere' regions present in Special Relativity.

According to the above the null velocity w is that of a pure rotation taking place at P and the 1-form of h is the corresponding angular momentum. The characteristic feature of the electromagnetic field, or the exterior derivative of the 1-form, is the frequency of rotation and not the field-magnitude. This field of frequencies is what figures in the Planck-Einstein relation.

F3 is likely to be the fundamental form of the Plank-Einstein relation. A simple insight into this possibility based on a 2-dimensional form of F3 is given in the Appendix D.

While (v, u) are the velocities of the wave aspect of matter at **P**, ($\overline{w}$, w) are the translational and rotational velocities of the corresponding field (g, h) at **P**. Unlike v and u which have the same magnitude and possess T-R unity, $\overline{w}$ and w have magnitudes 1 and 0 and possess T-R duality. In this case the T-R unity of (v, u) is the fulcrum on which the T-R duality ($\overline{w}$, w) of field balances.

The relative motion between v and u is independent of the purely hypothetical inertial frame and the motion of light that it engenders. In ET this inertial frame and the motion of light become replaced by the mutual referencing of v and u and the pair of null velocities w. These null velocities are the cause of the phenomenon of CMB radiation. Hence CMB radiation, which exists together with the dual congruence of (v, u) that satisfies the principle of relativity, is not in violation of the principle of relativity, as it appears to be at present, but is in conformity with it.

The state of balance between the geometrical opposites of point and line has finally produced their respective physical counterparts, matter and field; more precisely the wave aspect of matter that (v, u) produce and its corresponding field (g, h). Matter and field are also opposites which balance and in this case a state of unity of either matter or field is the fulcrum on which a state of duality of the other balances. Of these two cases of balance, the first has already taken place above. For T-R unity of the wave aspect of matter produced by (v, u) is the fulcrum on which the T-R duality of field (g, h) balances. The second case of a T-R unity of field acting as the fulcrum on which the T-R duality of the particle aspect of matter balances begins below.

4.2.2 Particle aspect of the wave-particle duality of matter

The sum and difference of v and u, namely $(v + u)/2$ ($\underline{v}$ = say) and $\pm(v - u)/2$ ($\underline{u}$ = say), transform the pair (v, u) into two alternative pairs $(\underline{v}, \underline{u})$ that correspond to the + and − signs of $\underline{u}$. Accordingly, F_1 that (v, u) satisfy transforms into the following two alternative pairs of equations $\underline{F}_1$ satisfied respectively by each of the two pairs $(\underline{v}, \underline{u})$.

$$(1-\beta)g_{ij}\underline{v}^j = \pm h_{ij}\underline{u}^j \qquad (17)$$

$$(1+\beta)g_{ij}\underline{u}^j = \pm h_{ij}\underline{v}^j \qquad (18)$$

$\underline{F}_1$ only produces squares of the eigen values β. Therefore both $+\beta$ and $-\beta$ are valid eigen values. Accordingly, when β changes sign $\underline{v}$ and $\underline{u}$ in $\underline{F}_1$ just interchange.

According to $\underline{F}_1$, each pair of $(\underline{v}, \underline{u})$ is orthogonal with respect to g indicating that each $(\underline{v}, \underline{u})$ is a pair of translational and rotational velocity vectors that possess T-R duality; hence each of the two pairs $(\underline{v}, \underline{u})$ represents a particle that possesses T-R duality. Each of these particles rotates about an axis that passes through the particle centre at which the translational and rotational velocities of (g, h) unify according to the following equation F_4 where b is a scalar parameter.

$$\{R_{ij} + C_{ij}\} = b\{g_{ij} \pm h_{ij}\}, \quad \det(g \pm h) = 0 \qquad (23)$$

Thus the field solution of F_2 becomes constrained so that $(g, +h)$ or $(g, -h)$ on the right-hand side of F_4 are those present at a particle centre.

The T-R unity of field $(g, +h)$ or $(g, -h)$ at a particle centre is the fulcrum on which the T-R duality of a particle balances as a whole. Loosely speaking the particle centre represents the particle as a whole. This representation is the cause of particle being viewed as a point in QM. But it is only a partial view of particle; hence QM is fundamentally incomplete.

The simplest form of the above two alternative particles is a sphere which rotates alternatively in opposite directions about an axis passing through its centre.

At present the two pairs ($\underline{v}$, $\underline{u}$) alternatively occupy the same spacetime manifold and then according to $\underline{F}_1$ they share the same g, but one is associated with only $+h$ and the other, only with $-h$. Hence, with respect to this spacetime manifold the two pairs ($\underline{v}$, $\underline{u}$) represent a particle and antiparticle as two alternative particle states. However, these two alternative particles are yet to acquire a state of balance as a pair. For this balance the particle centres, being fulcrums that are independent from those that they balance, are able to concurrently occupy a base manifold of their own characterised by an eigen field ($\underline{g}$, $\underline{h}$) which the following equations produce.

$$\{\underline{R}_{ij} + \underline{C}_{ij}\} = \underline{b}(\underline{g}_{ij} \pm \underline{h}_{ij}), \det(\underline{g} \pm \underline{h}) = 0 \qquad (24)$$

Just as in the case of h, $\underline{h}$ is also the sum of a 2-form and the exterior derivative of a 1-form. The base manifold equation (24) is a direct generalisation **(23)**. In a manner of speaking while **(23)** corresponds to the fulcrum of a balance, **(24)** corresponds to the body of the balance.

With respect to the base manifold characterised by ($\underline{g}$, $\underline{h}$) which (24) produces, the two pairs ($\underline{v}$, $\underline{u}$) represent two particles that possess their own spacetime manifolds and concurrently present at their own locations in the base manifold in a state balance with each other.

The solution of (24) necessitates that the base manifold splits into two separate manifolds with characteristic field equations $\mathbf{F_i}$ and $\mathbf{F_o}$ respectively (see §4.5.1) This separation is the result of the separation of the 2-form and the exterior derivative of the 1-form of $\underline{h}$. The two manifolds are referred to as the innermost and the outermost manifold of the two particles and the two particles share both these manifolds. More precisely, each particle centre, being a fulcrum of balance, and therefore being independent of the particle manifold whilst also co-existing with the particle manifold, occupies the innermost manifold. The outermost manifold which is the same as the empty spacetime manifold of GR externally envelops both particle manifolds.

The two particles that occupy the base manifold have their own spacetime manifolds characterised by their own tensor fields (*g*, *h*). The two elements (*g*, *h*) present at a particle centre, which are given by $\mathbf{F_4}$ are the same as ($\underline{g}$, $\underline{h}$) of the innermost manifold $\mathbf{F_i}$ present at the point which is occupied by the particle centre. Accordingly, at the particle centre the following set of conditions $\mathbf{F_5}$ is true.

$$g = \underline{g} \text{ and } h = \underline{h} \tag{43}$$

In F_5, while h is the sum of a 2-form and the exterior derivative of a 1-form of electromagnetic potentials, $\underline{h}$ is just a 2-form. Hence $h = \underline{h}$ unifies 1-form and 2-form of h at the particle centre as $\underline{h}$. As a result, the 1-form is now formally connected with the 2-form, or loosely speaking the electromagnetic field is now formally connected with the charge of the particle, at the particle centre which would also be the charge centre. The two electromagnetic fields of the two particles interact in the outermost manifold and this is the ET equivalent of the photon interchange between the two particles.

According to the foregoing F_4 and F_5 are particle-centre restrictions placed upon F_2 and F_i for each particle. Simultaneous particle-periphery restrictions occur for F_2 and F_0. Thus, each particle results from the simultaneous solution of five sets of equations F_2, F_i, F_4, F_5, and F_0. However, as the two particles share F_i, and F_0. the two particles and the base manifold that they occupy result from the simultaneous solution of only eight sets of equations. These eight sets are

(a) F_2, F_4 and F_5 for each of the two particles
(b) F'_i, and F'_0

Recall that §**4.2.1** began with a pair of mathematical points **P** and **P**$_1$ with disjoint neighbourhoods that occupy the n-dimensional spacetime R^n. The above two particles that occupy the base manifold are the physical counterparts of the above two mathematical points with disjoint neighbourhoods. The base manifold is the physical counterpart of the n-dimensional spacetime R^n. The two particle centres are the physical counterparts of the points **P** and **P**$_1$ and the rotations about axes through the particle centres are their disjoint neighbourhoods. Therefore, fundamentally the two particles also will not intersect however close they are in the base manifold and in addition they will also be fundamentally balanced as a pair on a fulcrum of optimum separation in the base manifold.

4.2.3 The basis of the physical world as a pair of balanced particles in the base manifold

Summarising the contents of **§4.2.1 and §4.2.2, in §4.2.1** the T-R unity of the wave aspect of matter is the fulcrum on which the T-R duality of field (g, h) balances. In §4.2.2 the T-R unity of (g, h) at the centre of a particle is the fulcrum on which the T-R duality of the particle manifold balances. This pair of balanced particles and the base manifold that they occupy are the basis of the physical world.

The simultaneous solution of the eight sets of equations places severe restrictions on the innermost and outermost manifolds and also on the two particle manifolds. Hence the occurrence of the central body system as the signature phenomenon of the physical world in which one particle is the central body and the other is a satellite. In the case of the central body system the central body and the satellite are structurally similar to those of the innermost and the outermost manifolds of the base manifold. In other words, the antisymmetric tensor of the central body consists virtually of the 2-form and that of the satellite consists virtually of the exterior derivative of the 1-form. Hence the central body virtually monopolises the base manifold and the satellite leaves the central body and the base manifold virtually unaffected by its presence. In this case the following are true.

(a) the satellite is of negligible mass
(b) the centres of the central body manifold and the base manifold approximately coincide
(c) the satellite moves in the gravitational fields of the central body and the base manifold whilst rotating about an axis that passes through the satellite centre.

The above two particles are the progenitors of the proton and the electron of the Hydrogen atom. Notice that as stated above the antisymmetric tensor of the electron virtually consists only of the exterior derivative of the 1-form similar to the accepted wisdom in standard physics. However, in ET it is only a pretty good approximation.

Base manifold of 'dark matter' occurs both locally and globally and is structurally capable of expansions and contractions. In the case of the central body system, the base manifold is local and envelops the complete central body system. The global base manifold which envelops the totality of visible matter is likely to be the cause of expansion and contraction of the physical world.

4.2.4 ET and the theories of Newton and Einstein

So far there had been two great theories of physics; those of Newton and Einstein. For both these, the physical world was finely tuned. For ET which subsumes both these theories the physical world is finely balanced. These three theories are completely free of uncertainties.

ET and Newton's physics share the same basis which is pairing. In Newton's physics action and reaction between a pair is equal and opposite. In ET a pair of opposites come together in a state of balance. The sets of equations F_1, F_2 and F_3 are the respective ET counterparts of the three laws of Newton (§5.4).

According to F_1 in Minkowski spacetime, v and u turn out to be Lorentzian boosts of each other. Thus, ET subsumes Special Relativity (SR) at the outset and it subsumes General Relativity (GR) in due course.

Therefore, in ET, theoretical physics may have completed a full circle and stands liberated high above the place on which it stood restricted at first.

4.3 Stage 1 of the Eigen Theory

4.3.1 Wave aspect of the wave-particle duality of matter

Appendix A develops the following set of equations F_1 satisfied by the pair of eigen velocity vectors (v, u) and their eigenvalue β.

$$\beta g_{ij} v^j = (g_{ij} + h_{ij}) u^j \qquad (A14)$$

$$\beta g_{ij} u^j = (g_{ij} - h_{ij}) v^j \qquad (A15)$$

In $\mathbf{R}^n$ the set of equations F_1 produces n pairs of (v, u) and n corresponding values of β as functions of g and h. The tensor h is the sum of a 2-form and the exterior derivative of a 1-form p, which means that h has the same number of elements as the tensor g. The elements of p are the counterparts of the elements of g that are arbitrary due to the Bianchi identities that g satisfies; hence, the elements of p are likewise primarily arbitrary. The tensors (g, h), being the same for each of the n pairs of eigen velocity vectors, are the kernel, or the substance, of matter. Therefore, in this introductory book on ET it suffices to focus on one pair of (v, u). Appendix A also shows that in Minkowski spacetime v and u are Lorentzian boosts of each other.

F_1 produces only squired values of β; hence β is positive or negative and these produce the following pairs of velocity vectors.

(a) $(\pm v, \pm u)$ as a reversal of both v and u also satisfies F_1
(b) $(\pm v, \mp u)$ as according to F_1 if β changes its sign then either v or u reverse

Fig. 1 shows sketches of the pairs of eigen velocity vectors in (a) as (s1, s2) and those in (b) as (s3, s4). In s1, the coordinate axes at the point of intersection of v and u are so configured that v and u are symmetrically placed with respect to the time-axis and co-planar with it. This coordinate configuration may be considered as a form of MCRF.

According to F_1 if v and u are interchanged then h reverses its sign. In s1 and s2 this interchange reverses the spatial direction of each of v and u. Accordingly, F_1 has CP symmetry since the sign of the 2-form component of h also corresponds to the sign of charge. In s3 and s4, if v and u are interchanged then time reverses for each of v and u; accordingly, F_1 has CT symmetry. If both v and u are reversed then time reverses for each of these; accordingly, F_1 has PT symmetry. Owing to the reversal of time, CT and PT symmetries are 'hidden'.

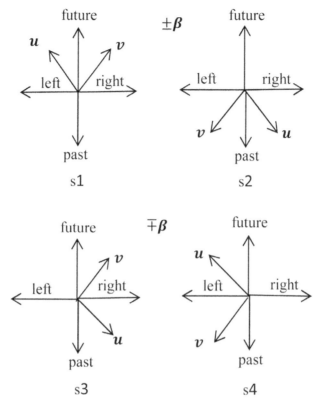

Fig. 1 The two pairs of solutions of F_1

4.3.2 The absolute conservation principle

Multiplying **(A14)** by *u* and **(A15)** by *v* and combining the results we get

$$\beta g_{ij} v^j u^i = g_{ij} v^j v^i = g_{ij} u^j u^i \qquad (1)$$

These together with F_1 produce the following.

$$(g_{ij} + \mathring{h}_{ij}) u^j v^i = 0 \qquad (2)$$

$$\check{h} = h/(1-\beta^2), \quad |\beta| \neq 1 \qquad (2a)$$

The scalars $g_{ij}u^j v^i$ and $\check{h}_{ij}u^j v^i$ in (2) are the ET counterparts of energy and angular momentum that are conserved separately in standard physics. According to (2), in ET the sum of these two counterparts is conserved as zero in perpetuity.

Thus, an absolute conservation principle of nothingness exists according to which the sum of energy and scalar angular momentum shared by the pair of eigen velocity vectors remains zero in perpetuity.

By analogy with Euclidean geometry, (1) leads to

$$|\beta| = \operatorname{sech} \xi \qquad (3)$$

where ξ is the hyperbolic angle between v and u. Accordingly, $|\beta|$ is confined to the closed interval [0, 1]. Then (1) indicates that

$$g_{ij} v^j u^i = \beta \qquad (4)$$

Equation (2) can now be re-written as

$$g_{ij} u^j v^i = -\check{h}_{ij} u^j v^i = \beta \qquad (5)$$

Hence, in ET the fundamental parameters of energy and scalar angular momentum of (v, u) unify as β. Finally, v and u are:

- (a) Of equal magnitudes, according to (1)
- (b) Orthogonal with respect to the general tensor $g + \check{h}$, according to (2)

(c) In a state of nothingness or nullity, according to (2)

(d) Such that for $|\boldsymbol{\beta}| = 1$ according to $\mathbf{F}_1$ they both become zero vectors which may be classed as singular as $\boldsymbol{\beta}$ is the magnitude of both v and u

According to the above features, (v, u) possess **E**quality, **O**rthogonality, **N**ullity and **S**ingularity, which bear the acronym EONS.

4.3.3 The transport equations of (v, u) and the field equation satisfied by (g, h)

As mentioned in §4.2.1, Appendix B produces a pair of mutual transport equations satisfied by (v, u) given below as **(B15)** and **(B16)**. In these equations, $v = d/ds$, $u = d'/ds$ where s is the path parameter.

$$du^k/ds + \{ij,k\}u^i v^j = 0 \qquad (B15)$$

$$d'v^k/ds + \{ij,k\}v^i u^j = 0 \qquad (B16)$$

$$\{ij,k\} = \{ij,k\}^s + \{ij,k\}^a \qquad (B17)$$

where

$$\{ij,k\}^s = \tfrac{1}{2} g^{lk}(\partial_j g_{il} + \partial_i g_{lj} - \partial_l g_{ji}) \qquad (B4)$$

$$\{ij,k\}^a = \tfrac{1}{2} h^{lk}(\partial_j h_{il} + \partial_i h_{lj} + \partial_l h_{ji}) \qquad (B9)$$

Equations **(B15)** and **(B16)** can be re-written as follows.

$$\left[\partial u^k/\partial x^j + \{ij,k\}u^i\right]v^j = 0 \qquad (6)$$

$$\left[\partial v^k/\partial x^j + \{ij,k\}v^i\right]u^j = 0 \qquad (7)$$

Since in *n*-spacetime (*g*, *h*) at present consists of $n^2 + n$ free variables, **(6)** and **(7)** can be reduced to

$$\partial u^k/\partial x^j + \{ij, k\} u^i = 0 \tag{8}$$

$$\partial v^k/\partial x^j + \{ij, k\} v^i = 0 \tag{9}$$

Equations **(8)** and **(9)** effectively represent just one equation as follows.

$$\partial \overline{w}^k/\partial x^j + \{ij, k\} \overline{w}^i = 0 \tag{10}$$

Differentiating **(10)** with respect to x^n we get

$$\partial^2 \overline{w}^k/\partial x^j \, \partial x^n + \partial\{ij, k\}/\partial x^n \, \overline{w}^i + \{ij, k\}\partial \overline{w}^i/\partial x^n = 0 \tag{10a}$$

We also have

$$\partial^2 \overline{w}^k/\partial x^j \, \partial x^n = \partial^2 \overline{w}^k/\partial x^n \, \partial x^j \tag{10b}$$

Equations **(10)**, **(10a)** and **(10b)** lead to the following set of equations F_2.

$$\{R^k_{ijn} + C^k_{ijn}\} \overline{w}^i = 0 \tag{11}$$

where

$$R^k_{ijn} = \{in, k\}^s_{,j} - \{ij, k\}^s_{,n} - \{mn, k\}^s \{ij, m\}^s \\ + \{mj, k\}^s \{in, m\}^s \tag{12}$$

$$C^k_{ijn} = \{in, k\}^a_{;j} - \{ij, k\}^a_{;n} - \{mn, k\}^a \{ij, m\}^a \\ + \{mj, k\}^a \{in, m\}^a \tag{13}$$

Since h is the sum of a 2-form and the exterior derivative of the 1-form p, F_2, which is antisymmetric in the indices j and n, consists of $(n^2 - n)\, n/2$ simultaneous equations in $(n^2 + 2n)$ unknowns; hence $n = 4$. Therefore, F_2 determines (g, h) together with a velocity $\overline{w}$ which is the translational velocity of the field (g, h) at **P** in 4-spacetime.

For the formulation of the field equation F_2 velocities u and v in (8) and (9) coincide and become the translational velocity $\overline{w}$ of (g, h) at **P**. Now β is both the magnitude of u and v and the hyperbolic secant of their angular separation. Hence in this case $|\beta| = 1$ and $\overline{w}$ is of unit magnitude. .

The following equation

$$\{R_{ij} + C_{ij}\}\overline{w}^i = 0, \quad \det(R + C) = 0 \qquad (14)$$

obtained by contracting F_2 with respect to k and n or k and j determines $\overline{w}$ at **P**.

According to SR, in flat spacetime symmetric metric tensor consists of 4 unit diagonal elements. The corresponding antisymmetric tensor is just the exterior derivative of the 4-element 1-form of the Maxwellean electromagnetic potentials. For these two forms of g and h, both R and C in F_2 vanish and F_2 is trivially satisfied. Accordingly, F_2 subsumes SR if the 1-form p that h contains is the set of Maxwellean electromagnetic potentials. A simple generalisation of Maxwell's electromagnetic equations in terms of h is given in Appendix C.

4.3.4 T-R unity of the wave aspect of matter and T-R duality of the corresponding field

With the field (g, h) that F_2 produces as input, F_1 determines an eigenvalue β and the corresponding pair of eigen velocity vectors (v, u) at **P**. For $|\boldsymbol{\beta}|$ within the open interval **(0, 1)** the vectors (v, u) transport each other and generate a dual congruence of world lines which possesses T-R unity. This dual congruence is the wave aspect of the wave-particle duality of matter. Equation **(14)** produces the translational velocity $\overline{w}$ of field (g, h) at **P**

For $\boldsymbol{\beta} = 0$, **(A14)** and **(A15)** produce the following set of equations that produces an alternative pair of rotational velocities w of (g, h) at **P**.

$$\{g_{ij} \pm h_{ij}\}w^i = 0, \quad \det(g \pm h) = 0 \qquad (14a)$$

The velocities $\overline{w}$ and w represent the T-R duality of (g, h).

4.4 Stage 2 of the Eigen Theory

4.4.1 Particle aspect of the wave-particle duality of matter

As described in §4.2.2 the pair of vectors (v, u), which possess T-R unity, transforms into two pairs of vectors $(\underline{v}, \underline{u})$, each of which possesses T-R duality as follows.

$$\underline{v} = (v + u)/2 \qquad (15)$$

$$\underline{u} = \pm(v - u)/2 \qquad (16)$$

The set of equations F_1 in turn transforms into a set F_1 as follows.

$$(1 - \beta)g_{ij}\underline{v}^j = \pm h_{ij}\underline{u}^j \qquad (17)$$

THE EIGEN THEORY OF THE PHYSICAL WORLD

$$(1+\beta)g_{ij}\underline{u}^j = \pm h_{ij}\underline{v}^j \tag{18}$$

According to these

$$g_{ij}\underline{v}^i\underline{u}^j = 0 \tag{19}$$

Therefore, each pair of ($\underline{v}$, $\underline{u}$) is translational and rotational and the two pairs split the wave aspect that v and u represent into two states of the particle aspect. These two particle states, or two particles for simplicity, corresponds to the + and − signs of the antisymmetric tensor h. In addition, since according to (19) each pair of ($\underline{v}$, $\underline{u}$) is orthogonal each particle possesses T-R duality.

Combining (17), (18) with (4) in §4.3.2 we get

$$\check{h}_{ij}\underline{v}^i\underline{u}^j = \mp\beta/2 \tag{20}$$

Finally (1) in §4.3.2 becomes

$$g_{ij}\underline{v}^i\underline{v}^j + g_{ij}\underline{u}^i\underline{u}^j = g_{ij}v^jv^i = g_{ij}u^ju^i \tag{21}$$

In the above system of equations both $+\beta$ and $-\beta$ are present on an equal footing.

If the set of mutual transport equations **(B15)** and **(B16)** satisfied by (v, u) is transformed into the set of equations satisfied by ($\underline{v}$, $\underline{u}$) then this set does not produce the set of field equations F_2 or an equivalent thereof. Therefore, each of the two pairs of ($\underline{v}$, $\underline{u}$), unlike (v, u), do not transport each other and generate a dual congruence of world lines. Each pair generates only a location in spacetime, which is the point of intersection of the pair of velocity vectors. Accordingly, the two pairs of ($\underline{v}$, $\underline{u}$) transform the dual congruence of line distribution to two pairs of point-distributions, which is a split of the wave aspect that (*v*, *u*) represent, into two particles. This point-distribution is the microscopic equivalent of the macroscopic point-like atomic-distributions that make up solid bodies.

4.5 Stage 3 of the Eigen Theory

4.5.1 The base manifold

§4.2.1 produces the following equation (24) that produces the tensor field ($\underline{g}$, $\underline{h}$) of the base manifold that the two particles occupy.

$$\{\underline{R}_{ij} + \underline{C}_{ij}\} = \underline{b}(\underline{g}_{ij} \pm \underline{h}_{ij}), \det(\underline{g} \pm \underline{h}) = 0 \qquad (24)$$

In (24) the number of elements of $\underline{g}$ and $\underline{h}$ present exceeds the number of component equations by 4. To rectify this mismatch (24) is first split into its symmetric and antisymmetric components as follows.

$$\underline{R}_{ij} + \underline{C}^s_{ij} = \underline{b}\underline{g}_{ij} \qquad (25)$$

$$\underline{C}^a_{ij} = \pm \underline{b}\underline{h}_{ij} \qquad (26)$$

Where

$$\underline{R}_{ij} = \{ik, k\}^s_{,j} - \{ij, k\}^s_{,k} - \{mk, k\}^s\{ij, m\}^s + \{mj, k\}^s\{ik, m\}^s \qquad (27)$$

$$\underline{C}^s_{ij} = \{ik,k\}^a_{;j} + \{mj,k\}^a \{ik,m\}^a \tag{28}$$

$$\underline{C}^a_{ij} = -\{mk,k\}^a \{ij,m\}^a - \{ij,k\}^a_{,k} \tag{29}$$

$$\{ij,k\}^s = \tfrac{1}{2}\underline{g}^{lk}(\partial_j\underline{g}_{il} + \partial_i\underline{g}_{lj} - \partial_l\underline{g}_{ji}) \tag{30}$$

$$\{ij,k\}^a = \tfrac{1}{2}\underline{h}^{lk}(\partial_j\underline{h}_{il} + \partial_i\underline{h}_{lj} + \partial_l\underline{h}_{ji}) \tag{31}$$

The 4 excess elements that equations (25) and (26) contain separate from the rest if the two components of $\underline{h}$, the 2-form and the exterior derivative of the 1-form p, separate. With this separation (25) and (26) split into two sets of equations that represent two separate manifolds, as follows.

Set 1: $\underline{h}$ consists only of the exterior derivative of the 1-form $\underline{p}$. Consequently $\underline{C}^s$ and $\underline{C}^a$ both vanish completely and (25) reduces to the following empty spacetime field equation in GR and 1-form $\underline{p}$ remains arbitrary.

$$\overline{\overline{R}}_{ij} = 0 \tag{32}$$

Equation (32) referred to as $\mathbf{F_o}$ produces the gravitational tensor field $\overline{\overline{g}}$ characteristic of the empty spacetime manifold that surrounds the two particles.

Set 2: Equations (25) and (26) are unchanged except that $\underline{h}$ consists only of the 2-form. Let these equations be numbered differently to avoid future confusion as follows:

$$\underline{R}_{ij} + \underline{C}^s_{ij} = \underline{b}\,\underline{g}_{ij} \qquad (25\text{m})$$

$$\underline{C}^a_{ij} = \pm\,\underline{b}\,\underline{h}_{ij} \qquad (26\text{m})$$

Equations **(25m)**, **(26m)** together with det $(\underline{g} \pm \underline{h}) = 0$, referred to as $\mathbf{F_i}$ produces the $(\underline{g}, \underline{h})$ tensor fields of a manifold that the centres of the two particles occupy. Due to the absence of the 1-form $\underline{p}$ of the Maxwellean electromagnetic potentials, this manifold is dark and its field content is the ET equivalent of dark matter.

While $\mathbf{F_o}$ is the structure of the empty spacetime manifold that envelops a particle as a whole externally, $\mathbf{F_i}$ is that of a completely opposite spacetime manifold as it envelops only the centre of a particle. Therefore, $\mathbf{F_o}$ and $\mathbf{F_i}$ are the field structures of the outermost and the innermost manifolds of the two particles, respectively which exists in addition to their own structures.

4.5.2 The complete structure of the two particles

As mentioned in **§4.2.2**, for each particle five sets of equations are involved which are $\mathbf{F_2}$, $\mathbf{F_i}$, $\mathbf{F_4}$, $\mathbf{F_5}$, **and** $\mathbf{F_0}$. Since the two particles share $\mathbf{F_i}$, and $\mathbf{F_o}$ the two particles and the innermost and outermost manifolds that they occupy are the simultaneous solution of eight sets of equations. The solution of these eight sets of equations provides in particular the particle centre positions in the innermost manifold and the corresponding 1-forms of momenta threat that $\mathbf{F_5}$ produces. While the particle centre positions are characteristic of the innermost manifold, the 1-forms thereat are characteristic of the manifolds of the two particles. These position and the 1-form of a particle are the ET counterparts of the position and momentum of a point-particle in standard physics.

4.5.3 The spherically symmetric cosmic central body system

If a particle is spherically symmetric then the particle centre is the same as the origin of the spherical polar coordinate system (t, r, θ, ϕ). If this spherically symmetric particle is the central body of a cosmic central body system, then a satellite would be a particle of negligible mass in comparison to the mass of the central body.

The satellite moves in the innermost and the outermost manifolds of the central body, outlined in **§4.5.1**. In accord with the spherical symmetry of the central body the satellite moves in a circle and both the innermost and the outermost manifolds simply contribute to it **(§6.3)**. In this context, notice that virtually all satellites in practice move in approximate circular paths around the central body. However, in general, the motion of the satellite would not be strictly circular and as a result, in the case of the simplest form of the central body system mentioned above, the satellite motion would be constrained as follows.

> (a) When the satellite is in close proximity to the central body, it effectively moves only in the outermost manifold. An example is the motion of a satellite in the solar system **(§6.3)**
>
> (b) When the satellite is sufficiently far away from the central body, it effectively moves only in the innermost manifold. An example is a satellite in a

galactic system, which is sufficiently far away from the galactic centre **(§6.3)**.

The occurrence of these constraints is due to the reduction of the satellite to a point. Otherwise, the combined solution of the eight sets of equations, mentioned in **§4.5.2** determines the motions of both the central body and the satellite.

A possibility exists that, just as in GR, the pair of particles may not have an analytic solution of their own.

CHAPTER 5

Simple applications of ET

5.1 The principle of uncertainty

In ET fundamentally a particle exists as one of a pair that occupy a base manifold in a state of balance that satisfies eight sets of equations. These equations in particular determine the two particle centre positions in the base manifold and their associated momenta 1-forms thereat. Therefore, in ET the position and the momentum of a particle by itself is unnatural. The principle of uncertainty in Quantum Mechanics is just an affirmation of this finding in ET.

5.2 Matter-antimatter

Whilst the pair of equations **(A14)** and **(A15)** represents the wave aspect of the particle, the following pair of equations represents the particle.

$$(1-\beta)g_{ij}\underline{v}^j = \pm h_{ij}\underline{u}^j \qquad (17)$$

$$(1+\beta)g_{ij}\underline{u}^j = \pm h_{ij}\underline{v}^j \qquad (18)$$

According to equations **(17)** and **(18)**, corresponding to the + or − signs of **h**, a particle consists of two alternatives rotating in opposite directions that share the same spacetime manifold. They possess the same mass but opposite charges. Since these are the properties of a pair of conventional matter and antimatter particles, in ET, a pair of matter and antimatter particles are two alternatives that are as the two sides of the proverbial coin.

These alternatives may co-exist in a suitable base manifold producing a pair of matter and antimatter particles.

5.3 The double-slit experiment

5.3.1 Particles of matter

In ET an isolated particle of matter is rationally and physically meaningless. Such a particle of matter may be created artificially as in a double slit experiment using the human ability to transcend physical limitations. However, because of its physically unrealistic character it may not behave in a rationally meaningful manner. However, if it is observed then it ceases to be isolated as it has then pair-formed with the observer, as a result, its behaviour then becomes rational. Such is the simple fundamental explanation of the strange results of the double-slit experiment in particle physics.

5.3.2 Photons

In ET an electromagnetic field of a particle of matter is a field of frequencies that satisfies the Planck-Einstein relation. Therefore, in ET the conventional photon, being a field of frequencies would go through both slits of the double slit experiment. This behaviour of the electromagnetic field accords with Planck's view on radiation-behaviour.

Since in ET a photon is a frequency field bizarre double-slit experimental may result under unusual experimental setups.

5.4 ET and Newtonian Mechanics

ET has three fundamental sets of equations F_1, F_2, and F_3. A one-to-one correspondence exists between these and the 3 laws of motion of Newton, as follows.

F_1 determines the primary motion as a pair of eigen velocities of T-R unity	First law states that the primary motion is a single uniform translational velocity
F_2 determines the force field that determines the primary motion.	Second law states the force that alters the primary motion
F_3 determines a pair of exchange photons	Third law states that forces occur in pairs that are equal and opposite.

The above comparison between ET and Newton's physics indicates that theoretical physics may now have journeyed a full circle and stands liberated high above the place on which she stood restricted.

CHAPTER 6

Application of ET to a cosmic central body system

6.1 Planetary and galactic systems

The simplest cosmic central body system is the one in which the central body and its innermost and outermost manifolds are static and spherically symmetric and the satellites are particles of masses negligible in comparison to the mass of the central body. In this case, the central body and its innermost manifold, share a common centre, which is at spatial rest.

Since the satellites are of negligible masses satellites move only in the outermost and innermost manifolds of the central body. Under static spherically symmetric conditions the metric characteristic of the innermost manifold is the exact opposite of the Schwarzschild metric characteristic of the outermost manifold of the central body.

6.2 The field solutions of the three manifolds of the central body.

6.2.1 Field solution of the outermost manifold of the central body

The solution of **(32)** in terms of spherical polar coordinates (t, r, θ, ϕ) is the following Schwarzschild metric. The parameter **m** is the length equivalent of the central body mass.

$$\overline{\overline{g}}_{ij} = diag\left\{\left(1 - \frac{2m}{r}\right), -1/\left(1 - \frac{2m}{r}\right), -r,^2 -r^2\sin^2\theta\right\} \quad (33)$$

6.2.2 Field solution of the innermost manifold of the central body

Appendix E produces the following solution of **(25m)** and **(26m)** in terms of spherical polar coordinates (t, r, θ, ϕ).

$$\underline{g}_{ij} = diag\left\{-\left(1 - \frac{r}{2M}\right), 1/\left(1 - \frac{r}{2M}\right), -r^2, -r^2\sin^2\theta\right\} \quad (E20)$$

$$\underline{h}_{ij} = \{\underline{h}_{01} = -\underline{h}_{10} = -1, \underline{h}_{23} = -\underline{h}_{32} = r^2\sin\theta\} \quad (E21)$$

The parameter **M** is the length equivalent of the 'dark matter' mass. Only the non-zero elements of $\underline{h}$ are shown in **(E21)**. At the manifold centre $\underline{g}$ is $(-1, 1, 0, 0)$ and in $\underline{h}$ only the two elements $\underline{h}_{01} = -\underline{h}_{10} = -1$ are non-zero.

6.2.3 Field solution of the central body manifold

This field, in general, is given by the following set of equations **F₂**.

$$\{R^k_{ijn} + C^k_{ijn}\}\overline{w^i} = 0 \qquad (11)$$

For the solution of the problem of satellite motion at hand the field solution of the central body manifold is superfluous.

6.3 Circular motion of a satellite of mass, negligible in comparison to the central body mass

Let *v* and *u* be the velocities of the satellite in the outermost manifold and the innermost manifold, respectively. In terms of spherical polar coordinates (t, r, θ, ϕ), these have the components $(v^0, 0, 0, v^3)$ and $(u^0, 0, 0, u^3)$, and they satisfy the following geodesic equations.

$$dv^k/ds + \{ij, k\}^s v^i v^j = 0 \qquad (34)$$

$$d'u^k/ds' + \{ij, k\}^s u^i u^j = 0 \qquad (35)$$

Symbols $\{ij, k\}^s$ and $\{ij, k\}^s$ are the respective connection coefficients that correspond to the two metrics **(33)** and **(E20)** in §6.2.1 and §6.2.2. Parameters *s* and *s'* in equations **(34)** and **(35)** are the respective path parameters. Since the motions are circular in both manifolds, only $k = 1$ in (34)) and (35) needs be considered and then these equations reduce to

$$s_v^2 = (rd\phi/dt)^2 = m/r \qquad (36)$$

$$s_u^2 = (rd'\phi/d't)^2 = r/(4M) \qquad (37)$$

where *r* is radial distance of the satellite, and s_v and s_u are the rotational speeds that correspond to the velocities *v* and *u* respectively.

Since the satellite motion is circular, s_v in **(36)** is the same as that given by the Newton's law of gravitation and m/r in **(36)** is the Newtonian gravitational potential at the radius r. If the density of the mass distribution in the innermost manifold is $1/(8\pi Mr)$ then s_u in **(37)** is also the same as that given by the Newton's law of gravitation and $r/(4M)$ is the Newtonian gravitational potential at the radius r. Accordingly, two Newtonian forces act radially inwards on the satellite and the magnitude of the speed s that corresponds to the resultant of these two forces is given by

$$|s| = \sqrt{m/r + r/4M} \qquad (38)$$

According to **(38)**, the contribution from the innermost manifold is negligible if

$$m/r \gg r/4M \qquad (39)$$

This in turn means

$$\underline{m}/\underline{r}^2 \gg 1 \qquad (40)$$

where $\underline{m}$ is the normalised mass of the central body and $\underline{r}$ is the normalised radial distance of the satellite, normalisations being with respect to mass M and the radius $2M$ respectively. For any reasonable estimate of M, the value of $\underline{m}/\underline{r}^2$ at the radial distance of the planet Pluto in the solar system is of the order of 10^5. Therefore, the innermost manifold makes no appreciable contribution to the dynamics of the solar system.

6.4 Use of equation (37) in §6.3 to obtain the contribution made by the innermost manifold to the rotational speeds of stars in the M33 galaxy

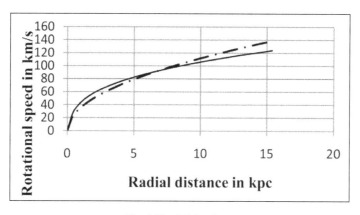

Fig. 2 The M33 galaxy

The value of *M* used for the computation of speeds shown in Fig.2 was obtained using **(37)** at the point, **6.833 kpc** and **92.35 km/s** of M33 galaxy data **(ref.3)**. This value of *M*, which is the total dark matter mass in the M33 galaxy is 7.47 x 10^{56} g.

In Fig. 2 speeds due to dark matter obtained from observational data on the M33 galaxy are shown as a continuous curve. Speeds that **(37)** produces are shown as a dashed curve.

6.5. Gravitational lensing

According to **(38)** in §6.3, for circular motion the total gravitational potential φ_r at a radius *r*, due to both the innermost manifold and the outermost manifold, is given by

$$\phi_r = m/r + r/4M \tag{41}$$

Accordingly, the angle of circular deviation of a ray of light due to both manifolds is given by

$$\alpha = 4(m/b + b/4M) \tag{42}$$

where *b* is the impact parameter.

Discussion and conclusions

ET has established that point particle as an entity by itself does not exist. ET also leads to the general conclusion that 'nothing exists on its own'. This has two meanings; (a) that 'nothingness' the mathematical equivalent of which is zero, exists on its own and (b) that 'no thing' exists on its own. According to ET a 'thing' only exists in a state of balance with another 'thing'. The best example in this case is either 'time' or 'space' as thing in itself. Time on its own is a single dimension which primarily exists as past, present and future which are completely different from each other. Space on its own consists of three dimensions which primarily are the same. Thus, it is clear that space and time are an archetypal pair of opposites and they primarily exist in a state of balance as spacetime.

The following are the principal features of ET.

1. According to ET the wave aspect of the wave-particle duality of matter produced by the mutual transport of a pair eigen velocity vectors holds the key to the complete structure of the physical world.

2. The sum of energy and a scalar angular momentum of the pair of eigen velocity vectors add up to zero in perpetuity. This is the only principle of conservation

in ET. Hence loosely speaking in ET only nothingness is conserved.

3. In standard physics, translational and rotational (T and R) motions are separate and so are the corresponding fields of gravitation and electromagnetism. In ET T–R unity and T- R duality are on an equal footing which implies that the gravitation and electromagnetism are also on an equal footing.

4. Gravitation and electromagnetism that grossly manifest as mass and charge of a particle of matter are in a constant state of interaction on an equal footing. As a result, the physical world is free of infinities of any kind and even those which apparently exist inside so-called black holes. *In standards physics these situations occur due to the state of separation between gravitation and electromagnetism.*

5. Spacetime is 4-dimensional. *In standard physics this dimensionality is assumed.*

6. A particle has an external spacetime manifold and an internal spacetime manifold, in addition to its own. *In standard physics only the external spacetime is present. At present the internal space time is wrongly recognised as dark matter.*

7. Translation of a particle takes place in the internal and external spacetime and the rotation takes place in its own spacetime. In standard physics translation takes place only in external spacetime and rotation is implicit.

8. Fundamentally matter and anti-matter are alternative forms of the same particle. *In standard physics they are completely separate.*

9. The above alternative forms manifest as the hadron and the lepton. *In standard physics matter and antimatter are assumed to be generated at the beginning of big bang in such a way that matter is in excess of antimatter and hence universe is composed mainly of matter.*

10. Eight sets of equations determine the complete structure of two interacting particles. These sets would simplify considerably in the case of the central body system which is the signature of the physical world.

11. M33 data verifies the accuracy of the structure of the so-called dark matter.

Appendix A

Set of algebraic equations F_1 that the pair of eigen velocity vectors (*v*, *u*) and its eigenvalue β satisfy

Consider two vectors (*v*, *u*) that relate to each other in general as

$$bv^i = a^i_k u^k \tag{A1}$$

$$b'u^i = a'^i_k v^k \tag{A2}$$

where *a* and *a'* are tensors of type (1, 1) and *b* and *b'* are two scalar variables. Rewriting **(A1)** and **(A2)** with *v* and *u* in their coordinate form we get

$$b dx^i/ds = a^i_k d'x^k/ds' \tag{A3}$$

$$b' d'x^i/ds' = a'^i_k dx^k/ds \tag{A4}$$

where x^0 is the time coordinate and *s* and *s'* are path parameters. Let *b*, *b'* and the two path parameters be adjusted so that

$$b(ds'/ds) = b'(ds/ds') = \beta \tag{A5}$$

where β is a scalar variable. Then **(A1)** and **(A2)** become

$$\beta v^i = a^i_k u^k \qquad \text{(A6)}$$

$$\beta u^i = a'^i_k v^k \qquad \text{(A7)}$$

where

$$v^i = dx^i/ds \qquad \text{(A8)}$$

$$u^i = d'x^i/ds \qquad \text{(A9)}$$

The path parameter s is the ET equivalent of proper time in Relativity.

Equations **(A6)** and **(A7)** can be re-written as follows.

$$\beta g_{ij} v^j = (g_{ij} + h_{ij}) u^j \qquad \text{(A10)}$$

$$\beta g'_{ij} u^j = (g'_{ij} + h'_{ij}) v^j \qquad \text{(A11)}$$

where g and g' are symmetric and h and h' are antisymmetric. The condition that v and u are eigen vectors is a' in **(A7)** is the transpose of a in **(A6)** (**ref.1**). This condition is satisfied if

$$g' = g \qquad \text{(A12)}$$

$$h' = -h \qquad \text{(A13)}$$

Re-writing **(A10)** and **(A11)** to include **(A12)** and **(A13)** we have

$$\beta g_{ij} v^j = (g_{ij} + h_{ij}) u^j \qquad (A14)$$

$$\beta g_{ij} u^j = (g_{ij} - h_{ij}) v^j \qquad (A15)$$

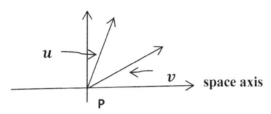

FIG. A1 Time and space axes coplanar with (v, u)

In Minkowski spacetime let the coordinate axes at the point of intersection of v and u be so configured that v and u are coplanar with the time-axis and a space-axis, as shown in Fig. A1.

In this case, **(A14)** and **(A15)** produce the following.

$$\sqrt{1-s^2}\begin{bmatrix} v^0 \\ v^1 \end{bmatrix} = \begin{bmatrix} 1 & s \\ s & 1 \end{bmatrix}\begin{bmatrix} u^0 \\ u^1 \end{bmatrix} \qquad (A16)$$

$$\sqrt{1-s^2}\begin{bmatrix} u^0 \\ u^1 \end{bmatrix} = \begin{bmatrix} 1 & -s \\ -s & 1 \end{bmatrix}\begin{bmatrix} v^0 \\ v^1 \end{bmatrix} \qquad (A17)$$

$$s = h_{01} \qquad (A18)$$

According to these the two eigen velocity vectors v and u are Lorentzian boosts of each other with a boost speed of s.

Appendix B

The mutual transport equations of (v, u)

Let an arbitrary infinitesimal perturbation δx^i, $i = 0,...n-1$, be applied to the point of intersection of the pair of eigen velocity vectors (v, u) so that β is kept invariant. Then according equations **(5)** in §**4.3.2** we have

$$\delta(g_{ij}v^i u^j) = \delta\beta = 0 \tag{B1}$$

$$\delta(h_{ij}v^i u^j) = \delta\beta(1 - \beta^2) = 0 \tag{B2}$$

On expanding **(B1)** we get

$$[g_{kn}d/ds\left\{(d'x^n/ds)\right\} + g_{kn}d'/ds\left\{(dx^n/ds)\right\}$$
$$+ 2g_{ko}\{ij, o\}^s (d'x^i/ds)\ (dx^j/ds)]\,\delta x^k = \beta_g \tag{B3}$$

where

$$\{ij, k\}^s = \frac{1}{2}g^{lk}\left((\partial g_{il}/\partial x^j) + (\partial g_{lj}/\partial x^i) - (\partial g_{ji}/\partial x^l)\right) \tag{B4}$$

$$\beta_g = d/ds\{g_{ij}\delta x^i(d'x^j/ds)\} + d'/ds\{g_{ij}(dx^i/ds)\delta x^j\} \tag{B5}$$

Since δx^i is arbitrary, **(B3)** effectively represents n equations, which respectively correspond to the n elements of δx.

Let δx^i be an element of δx with the remaining elements of δx set to zero. Let the scalars within the curly brackets in **(B5)** be denoted by c'^l and c^l as follows.

$$g_{lj}\delta x^l (d'x^j/ds) = c'^l \qquad \text{(B5a)}$$

$$g_{lj}\delta x^l (dx^j/ds) = c^l \qquad \text{(B5b)}$$

The summation on l is suspended in **(B5a)** and **(B5b)**. Eliminating δx^l from **(B5a)** and **(B5b)** we get

$$(g_{lj}/c'^l)(d'x^j/ds) = (g_{lj}/c^l)(dx^j/ds) \qquad \text{(B5c)}$$

In the n independent equations that correspond to $l = 0,\ldots n-1$, which **(B5c)** represents, the ratios between c'^l and c^l can be maintained constant using the set of n arbitrary elements of g. Then because x^j is arbitrary, c'^l and c^l can be individually maintained constant. Therefore

$$\beta_g = 0 \qquad \text{(B6)}$$

Then we have

$$[g_{kn}d/ds\{(d'x^n/ds)\} + g_{kn}d'/ds\{(dx^n/ds)\}$$
$$+ 2g_{ko}\{ij,o\}^s (d'x^i/ds)(dx^j/ds)] = 0 \qquad \text{(B7)}$$

On expanding **(B2)** in the same way that **(B1)** was expanded, we get

$$[h_{kn} d/ds \{(d'x^n/ds)\} - h_{kn} d'/ds \{(dx^n/ds)\}$$
$$+ 2h_{ko} \{ij, o\}^a (d'x^i/ds)(dx^j/ds)] \delta x^k = \beta_h \quad \text{(B8)}$$

Where

$$\{ij, k\}^a = \frac{1}{2} h^{lk} \left((\partial h_{il}/\partial x^j) + (\partial h_{lj}/\partial x^i) + (\partial h_{ji})/(\partial x^l) \right) \text{(B9)}$$

$$\beta_h = d/ds \{h_{ij} \delta x^i (d'x^j/ds)\} + d'/ds \{h_{ij} (dx^i/ds) \delta x^j\} \quad \text{(B10)}$$

The reasoning that was used to obtain **(B6)** is applicable in this case also as the 1-form that **h** contains consists of **n** arbitrary elements. Thus, we get

$$\beta_h = 0 \quad \text{(B11)}$$

$$[h_{kn} d/ds \{(d'x^n/ds)\} - h_{kn} d'/ds \{(dx^n/ds)\}$$
$$+ 2h_{ko} \{ij, o\}^a (d'x^i/ds)(dx^j/ds)] = 0 \quad \text{(B12)}$$

Equations **(B7)** and **(B12)** can be re-written as

$$d/ds \{d'x^k/ds\} + d'/ds \{dx^k/ds\}$$
$$+ 2\{ij, k\}^s (d' x^i/ds)(dx^j/ds) = 0 \quad \text{(B13)}$$

$$d/ds \{d'x^k/ds\} - d'/ds \{dx^k/ds\}$$
$$+ 2\{ij, k\}^a (d'x^i/ds)(dx^j/ds) = 0 \quad \text{(B14)}$$

Adding **(B13)** and **(B14)** and subtracting **(B14)** from **(B13)** we get the following equations that determine the mutual transport of (**v**, **u**).

$$d/ds\{d'x^k/ds\} + \{ij,k\}(d'x^i/ds)(dx^j/ds) = 0 \qquad \textbf{(B15)}$$

$$d'/ds\{dx^k/ds\} + \{ij,k\}(dx^i/ds)(d'x^j/ds) = 0 \qquad \textbf{(B16)}$$

where

$$\{ij,k\} = \{ij,k\}^s + \{ij,k\}^a \qquad \textbf{(B17)}$$

Appendix C

A simple generalisation of Maxwell's electromagnetic equations

In ET, the antisymmetric tensor h is the sum of a 2-form and the exterior derivative of a 1-form p. This h can be used to generalise the following Maxwell's electromagnetic equations

$$F_{ij} = \partial_j p_i - \partial_i p_j \tag{C1}$$

$$\partial_j \{\sqrt{-g}\, g^{im} g^{jn} F_{mn}\} = \sqrt{-g}\, J^i \tag{C2}$$

where J is the charge-current density vector. If the 2-form component of h, say $\underline{h}$, is assumed to be related to J as

$$\partial_j \{\sqrt{-g}\, g^{im} g^{jn} \underline{h}_{mn}\} = -\sqrt{-g}\, J^i \tag{C3}$$

then in **ET** the above Maxwell's equations become generalised as

$$\partial_j \{\sqrt{-g}\, g^{im} g^{jn} h_{mn}\} = 0 \tag{C4}$$

Appendix D

A simple insight into Planck-Einstein Relation

In 2-dimensional Minkowski spacetime, equation

$$\{g_{ij} \pm h_{ij}\}w^i = 0, \quad \det(g \pm h) = 0 \tag{14b}$$

reduces to just the following.

$$dt/ds = dx/ds \tag{D1}$$

where dt and dx are the time and space increments of the photon travel and ds is path parameter increment yet to be established owing to the null character of w. Now dt/ds is the dimensionless photon energy E'. Therefore, (D1) becomes

$$E' = dx/ds \tag{D2}$$

Since this energy E' is dimensionless it has to be multiplied by a fundamental energy constant to obtain E in the usual units of energy. This constant is Planck energy $(\hbar c/G)^{1/2}c^2$ where $\hbar$ is Planck's reduced constant, c is speed of light and G is gravitation constant. Then (D2) becomes

$$E = (\hbar c/G)^{1/2} c^2 (dx/ds) \tag{D3}$$

Now **(22)** is the result of unification of translational and rotational motions of light. This unification in this 2-dimensional case connects dx with an incremental hyperbolic angle $d\xi$ using Planck distance as follows.

$$dx = (Planck\ distance)\,d\xi \qquad \textbf{(D4)}$$

Since Planck distance is $(\hbar G/c^3)^{1/2}$, **(D3)** becomes

$$E = (\hbar c)(d\xi/ds) \qquad \textbf{(D5)}$$

If the normal angular velocity is the same as $d\xi/(ds/c)$ then that would define the path parameter s and **(D5)** becomes the same as Planck-Einstein relation.

$$E = h\nu \qquad \textbf{(D6)}$$

where ν is normal frequency.

Appendix E

**Solution of (25m) and (26m) in §4.5.1,
for static spherically symmetric space**

Equations **(25m)**, **(26m)** and their auxiliary equations are as follows.

$$\underline{R}_{ij} + \underline{C}^s_{ij} = \underline{b}\,\underline{g}_{ij} \qquad (25m)$$

$$\underline{C}^a_{ij} = \pm\,\underline{b}\,\underline{g}_{ij} \qquad (26m)$$

where

$$\underline{R}_{ij} = \{ik,k\}^s_{,j} - \{ij,k\}^s_{,k} - \{mk,k\}^s\{ij,m\}^s + \{mj,k\}^s\{ik,m\}^s \qquad (27)$$

$$\underline{C}^s_{ij} = \{ik,k\}^a_{;j} + \{mj,k\}^a\{ik,m\}^a \qquad (28)$$

$$\underline{C}^a_{ij} = -\{mk,k\}^a\{ij,m\}^a - \{ij,k\}^a_{,k} \qquad (29)$$

$$\{ij,k\}^s = \tfrac{1}{2}\underline{g}^{lk}\left(\partial_j \underline{g}_{il} + \partial_i \underline{g}_{lj} - \partial_l \underline{g}_{ji}\right) \qquad (30)$$

$$\{ij,k\}^a = \tfrac{1}{2}\underline{h}^{lk}\left(\partial_j \underline{h}_{il} + \partial_i \underline{h}_{lj} + \partial_l \underline{h}_{ji}\right) \qquad (31)$$

For static spherically symmetric conditions $\underline{g}$ and $\underline{h}$ in spherical polar coordinates (t, r, θ, ϕ) have the following forms.

$$\underline{g}_{ij} = diag\{e^\nu, -e^\lambda, -r^2, -r^2\sin^2\theta\} \qquad \text{(E1)}$$

$$\underline{h}_{ij} = \{\underline{h}_{01} = -\underline{h}_{10} = -e^\alpha, \underline{h}_{23} = -\underline{h}_{32} = \sin\theta e^\rho\} \qquad \text{(E2)}$$

Parameters ν, λ, α and ρ are functions of r only and all other elements of $\underline{h}$ are zero. The elements of $\{ij, k\}^s$ that correspond to $\underline{g}$ in (E1) are well established in the literature and they can be found on page 84 of ref. 2. The non-zero elements of $\{ij, k\}^a$ that correspond to $\underline{h}$ in (E2) are as follows.

$$\{12, 2\}^a = -\rho'/2 \qquad \text{(E3)}$$

$$\{13, 3\}^a = -\rho'/2 \qquad \text{(E4)}$$

$$\{21, 2\}^a = +\rho'/2 \qquad \text{(E5)}$$

$$\{23, 0\}^a = +\sin\theta e^{\rho-\alpha}\rho'/2 \qquad \text{(E6)}$$

$$\{31, 3\}^a = +\rho'/2 \qquad \text{(E7)}$$

$$\{32, 0\}^a = -\sin\theta e^{\rho-\alpha}\rho'/2 \qquad \text{(E8)}$$

The accent,', on a symbol denotes differentiation with respect to r. On substituting these elements of $\{ij, k\}^a$ in the expression for $\underline{C}^a_{ij}$ in (29), we get

$$\underline{C}^a_{ij} = 0 \qquad \text{(E9)}$$

Hence, it follows that the parameter $\underline{b}$ in (25m) and (26m) is zero. On substituting the above elements of $\{ij, k\}^a$ in the expression for $\underline{C}^s_{ij}$ in (28), we get the following.

$$\underline{C}^s_{00} = e^{v-\lambda}v'(\rho')/2 \tag{E10}$$

$$\underline{C}^s_{11} = -(\rho'') + \lambda'(\rho')/2 - (\rho')^2/2 \tag{E11}$$

$$\underline{C}^s_{22} = -re^{-\lambda}(\rho') \tag{E12}$$

$$\underline{C}^s_{33} = -r\sin^2\theta e^{-\lambda}(\rho') \tag{E13}$$

Elements $\{ij, k\}^s$ of the tensor $\underline{R}_{ij}$ in (27) can be found on page **85** of ref. 2 where R has been denoted as G. On substituting in **(25m)** these elements $\{ij, k\}^s$ and the above elements of, $\underline{C}^s_{ij}$ we get

$$e^{v-\lambda}\left(-\frac{1}{2}v'' + \frac{1}{4}\lambda'v' - \frac{1}{4}v'^2 - \frac{v'}{r}\right) + \frac{1}{2}e^{v-\lambda}v'\rho' = 0 \tag{E14}$$

$$\frac{1}{2}v'' - \frac{1}{4}\lambda'v' + \frac{1}{4}v'^2 - \frac{\lambda'}{r} - (\rho'') + \frac{1}{2}\lambda'(\rho') - \frac{1}{2}(\rho')^2 = 0 \tag{E15}$$

$$e^{-\lambda}\left(1 + \frac{1}{2}r(v' - \lambda')\right) - 1 - re^{-\lambda}(\rho') = 0 \tag{E16}$$

The parameters v, λ and ρ that satisfy equations **(E14)** to **(E16)** are as follows.

$$\rho' = 2/r \tag{E17}$$

$$e^v = -(1 - r/(2M)) \tag{E18}$$

$$e^\lambda = -1/(1 - r/(2M)) \tag{E19}$$

M in (E18) and (E19) is a constant of integration. (E17) to (E19) together with the condition $\det(\underline{g} \pm \underline{h}) = 0$ produces $\underline{g}$ and $\underline{h}$ in (E1) and (E2), as follows.

$$\underline{g}_{ij} = diag\{-\left(1-\frac{r}{2M}\right), 1/\left(1-\frac{r}{2M}\right), -r^2, -r^2 sin^2\theta\} \quad \text{(E20)}$$

$$\underline{h}_{ij} = \{\underline{h}_{01} = -\underline{h}_{10} = -1,$$
$$\underline{h}_{23} = -\underline{h}_{32} = kr^2 sin\theta\} \quad \text{(E21)}$$

where **k** is a constant which may be set to unity.

Appendix F

Gravitational redshift due to spacetime curvature of the innermost manifold

The metric tensor field of the innermost manifold, obtained in Appendix E, for static spherically symmetric space is the following.

$$\underline{g}_{ij} = diag\{-\left(1-\frac{r}{2M}\right), 1/\left(1-\frac{r}{2M}\right), -r^2, -r^2 sin^2\theta\} \quad \text{(E20)}$$

Let a photon be emitted at a distance r from the origin O of the system of spherical polar coordinates (t, r, θ, ϕ), and let its frequency at this point of emission be v_e. On reaching O let the photon frequency become v_r. These two frequencies relate to each other as (**ref.4**)

$$v_e/v_r = \left(1 - r/(2M)\right)^{-1/2} \quad \text{(F1)}$$

This frequency ratio in terms of a recessional speed s, is given by

$$v_e/v_r = \sqrt{(1+s)/(1-s)} \quad \text{(F2)}$$

Combining **(F1)** and **(F2)**, we get

$$s = (r/(2M))/(2 - r/(2M)) \qquad \text{(F3)}$$

For comparing with Hubble's law, let **(F3)** be re-written as $s = \underline{H}r$, where

$$\underline{H} = 1/(4M - r) \qquad \text{(F4)}$$

According to **(F4)** the maximum value of $\underline{H}$ is $1/(2M)$ that occurs at $r = 2M$. Now M has been estimated in §6.4 as **7.47 x 10^{56} g** which works out to **1.799x10^{07}** kpc. Hence the maximum value of $\underline{H}$ (=1/(2M)) is **8.34 kms^{-1}Mpc^{-1}**. This value of $\underline{H}$ is considerably less than the present value of the Hubble's constant, **73.8 kms^{-1}Mpc^{-1}**. Therefore, the gravitational curvature of the innermost manifold has only a small effect on the observed Hubble expansion of the physical universe.

REFERENCES

[1] George Arfken, 1985, *Geometrical Methods for Physicists,* (Academic Press Inc., 1985), p.235.

[2] Eddington, A. S, *'The Mathematical Theory of Relativity'*, Chelsea Publishing Company, New York, N. Y. Third Edition, 1975, P 84-85

[3] Corbelli, E and Salucci, P, 2000, 'The Extended Rotation Curve and the Dark Matter Halo of M33', *Mon. Not R. Astron Soc.* 311, 441- 447

[4] Schutz, B. F, *'A First Course in General Relativity'*, Cambridge University Press, 1985, P 255

[5] Schutz, B. F, *'Geometrical methods of mathematical physics'*, Cambridge University Press, 1987, P3

BIBLIOGRAPHY

1. R K Adair. The Great Design, Oxford University Press, 1987
2. M Born. Atomic Physics, Blackie & Son Limited,1972
3. F Close. The Cosmic Onion, Heinemann Educational Books, 1985
4. M Crampin and F A E Pirani. Applicable Differential Geometry, Cambridge University Press, 1988
5. P Davies et al. The New Physics, Cambridge University Press, 1990
6. A. d'abro. The Rise of the New Physics, Vols. One and Two, Dover Publications, Inc., New York, 1951
7. A S Eddington. The Mathematical Theory of Relativity, Chelsea Publishing Company, N Y, 1975
8. A Einstein. The Meaning of Relativity, Methuen & Co. Ltd., 1922
9. R P Feynman. Lectures on Physics, Vols. I, II and III, Addison-Wesley Publishing Company, Massachusetts, 1964
10. R P Feynman. Quantum Electro-dynamics, The Benjamin/Cummings Publishing Company, Inc., Massachusetts, 1983
11. R P Feynman and S Weinberg. Elementary Particles and the Laws of Physics, (The 1986 Dirac Memorial Lectures), Cambridge University Press, 1989

12. R P Feynman. The Character of Physical Law, The M.I.T. Press, Massachusetts, 1983
13. S W Hawking. A Brief History of Time, Bantam Press, London, 1988
14. C Lanczos. The Variational Principles of Mechanics, University of Toronto Press, Toronto, 1970
15. H A Lorentz et al. The Principle of Relativity, Dover Publications, Inc., 1952
16. C W Misner et al. Gravitation, W H Freeman and Company, San Francisco, 1973
17. A Pais. 'Subtle is the Lord...' (The Science and the Life of Albert Einstein), Oxford University Press, 1982
18. R Penrose. The Emperor's New Mind, Vintage, 1989
19. E Schrodinger. Space-Time Structure, Cambridge University Press, 1988
20. B F Schutz. A First Course in General Relativity, Cambridge University Press, 1985
21. B F Schutz. Geometrical Methods of Mathematical Physics, Cambridge University Press, 1987
22. H Weyl. Space Time Matter, Dover Publications, Inc., 1952

Lightning Source UK Ltd.
Milton Keynes UK
UKHW041051081022
410034UK00017B/196